Praise for *The Spiritual Nature of Animals*

"Dr. Karlene Stange weaves a heartwarming web composed of her personal experiences as a veterinarian caring for all our kindred spirits, along with insights from ancient wisdom traditions and perspectives on our boundless interconnectedness with all beings. As a veterinarian who has shared my own journey of awakening through caring for our animal friends, I have thoroughly enjoyed reading Stange's story and know that it will benefit countless beings! Allow it to touch your heart and spirit, and enjoy!"
— **Allen M. Schoen, DVM, MS, PhD (hon.),** author of *Kindred Spirits*

"In her book, Karlene Stange addresses some of the hardest questions veterinarians face and does so with a fascinating and surprisingly seamless blend of science/medicine, spirituality, and personal experience. She gives us the kernels of truth about the spirit, or *anima*, of animals from the perspective of each major religion and distills these into a very practical philosophy that I think is summed up in the last paragraph of her book: 'We must be grateful for the service and friendship animals provide, and celebrate their lives, honor loving memories, and have no regrets.' This book is well researched and passionate and comes straight from the heart of someone who loves animals and has committed her life to their health and well-being."
— **Susan J. Tornquist, DVM, MS, PhD, Dip ACVP,**
Lois Bates Acheson Dean, College of Veterinary Medicine,
Oregon State University

"All too often, humans have difficulty understanding the spiritual component of animals. Based on twenty years of research, Dr. Karlene Stange shares wisdom from many cultures and from the creatures themselves to help us better understand. Everyone who cares about animals will benefit from this book."
— **Huisheng Xie, DVM, PhD,** founder and dean,
Chi Institute of Traditional Chinese Veterinary Medicine (www.tcvm.com)

"Humankind's eternal quest for the meaning of life and death has spawned innumerable religions and spiritual teachings. Dr. Karlene Stange brings us along on her journey of exploration into varied religious and cultural precepts and how they intertwine with the animal world. She weaves stories of her everyday life as a veterinarian into the pursuit of understanding the intimate relationship between people and animals. Along this journey, she continually seeks ways that she, as a healer, can positively affect this spiritual bond. In this book, you will find stories that make you laugh, others that bring tears to your eyes, and still others that evoke enlightening views about the world we cohabit with animals and how this affects our belief systems. I highly recommend this inspiring book."

— **Nancy S. Loving, DVM,** author of
All Horse Systems Go and *Go the Distance*

"In *The Spiritual Nature of Animals*, Karlene Stange, DVM, finally gives the issue of animal spirituality the attention it deserves. Expertly written with fascinating stories of her experiences as a country veterinarian, her book covers every conceivable aspect of religion and animals, death and the afterlife for animals, and how animals mirror their people. Not only is this work a fascinating look into the private world of those who care for our animals, but it is the definitive examination of animals and spirituality. And along the incredible journey of reading it you will also learn how to live your life without pain and judgment."

— **Marta Williams,** author of *Learning Their Language*

The Spiritual Nature of
ANIMALS

The Spiritual Nature of
ANIMALS

*A Country Vet Explores the Wisdom,
Compassion, and Souls of Animals*

Karlene Stange, DVM

New World Library
Novato, California

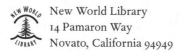

New World Library
14 Pamaron Way
Novato, California 94949

Text design by Megan Colman
Typography by Tona Pearce Myers

Library of Congress Cataloging-in-Publication Data
Names: Stange, Karlene, [date], author.
Title: The spiritual nature of animals : a country vet explores the wisdom, compassion, and
    souls of animals / Karlene Stange, DVM.
Description: Novato, California : New World Library, 2017. | Includes bibliographical references and index.
Identifiers: LCCN 2017025307 (print) | LCCN 2017041298 (ebook) | ISBN 9781608685165
    (Ebook) | ISBN 9781608685158 (alk. paper)
Subjects: LCSH: Animal psychology. | Animals—Religious aspects. | Human-animal relationships. | Stange, Karlene, [date] | Veterinary medicine—Anecdotes.
Classification: LCC QL785 (ebook) | LCC QL785 .S815 2017 (print) | DDC 591.5—dc23
LC record available at https://lccn.loc.gov/2017025307

First printing, November 2017
ISBN 978-1-60868-515-8
Ebook ISBN 978-1-60868-516-5

Printed in Canada on 100% postconsumer-waste recycled paper

New World Library is proud to be a Gold Certified Environmentally Responsible Publisher. Publisher certification awarded by Green Press Initiative. www.greenpressinitiative.org

10   9   8   7   6   5   4   3   2   1

Contents

CHAPTER 1

Ambulatory Horse Doctor

"Honey, if you don't come, she's going to die," said Mr. Hall.

The day's schedule was full. My neck ached. Instead of taking lunch, I was driving to the chiropractor.

"She fell in the irrigation ditch this morning, and now she's too weak to stand," explained Mr. Hall about his three-day-old foal.

My psychotherapist warned me. She said, "You need to take care of yourself. Take more time off; just say no." That was easy for her to say.

"I'll come right now," I said, turning the truck around. A three-day-old foal is a neonate — a term containing deep meaning for a veterinarian. People had planned for and anticipated this foal's arrival for more than a year; the mare had spent 345 days in gestation to produce this precious, fragile being full of hope and promise; and now, after three days, it was ready to drop dead at the snap of a finger. The idea of a sick neonatal foal set off alarm bells in my brain and made my heart race.

It was June 6, 1995, and a heavy morning frost was making life challenging for a newborn. This nice old man, Mr. Hall, was right — she would die without help. My neck could wait, and if I had any snacks, I would eat them in the truck.

From 1990 to 2010, I worked as an ambulatory veterinarian, and I lived in my truck. Four trucks met their mileage limits during that time, but I always drove white three-quarter-ton F250 Fords, four-wheel-drive vehicles with a regular cab and full-length running boards that allowed all five foot four inches of me to reach into the veterinary utility box that lined the eight-foot bed. That utility box, called a Porta-Vet, contained a hidden water tank in the center. The side compartments raised up at a slant, exposing a deep floor near the cab and shelves holding medications and tools toward the rear. On the driver's side, the deep area closest to the cab held a refrigerator where I kept vaccines, hormones, and antibiotics. Overhead, a roll of paper towels hung from the door. On the floor sat a trash can and a lockbox for controlled substances, such as euthanasia solution, analgesics, and anesthetics. Just above and to the right of the refrigerator sat a covered dish filled with surgical tools: needle holders, scissors, curved needles, hemostats, and forceps. Another slotted compartment farther to the rear contained hypodermic needles, acupuncture needles, syringes, a stethoscope, thermometer, and test tubes for blood collection. On the passenger side, in the section of the Porta-Vet closest to the cab, was a deep compartment where I kept an X-ray machine and a case of intravenous fluid bags. Toward the back were shelves with hoof testers, hoof nippers, an oral speculum, dental floats, deworming medications, and oral anti-inflammatory drugs.

The tailgate of the utility box dropped to provide a work space. The left inside wall of the rear compartment was equipped with switches for lights and a water pump. A hose to the water tank hung next to them on a bracket. A large pull-out drawer held a stainless-steel bucket and a box of ropes and halters; on the right, a small upper drawer contained bandage materials, while a lower drawer transported a computer and printer, each enclosed in padded gun cases.

The truck's cab interior was always gray. The driver's side door held sunscreen, hand sanitizer, and snack bars — or so I hoped that June morning. Jammed behind the seat were coveralls, boots, down jackets, vests, and numerous hats and gloves to suit the ever-changing weather. In the early-model trucks, the cellular phone was perched on a post in the center of the floor. (When the first cellular phone company came to town, I won that phone in a contest by writing in fifty words or less why I needed a cellular phone.) Mounted on the hood of each truck was a bronze horse-head hood ornament with a liver-chestnut patina.

My truck was more than a mobile animal surgical office. It was my home away from home.

Mr. Hall's filly was as cold as the weather. Because neonatal horses do not have the ability to thermoregulate like adults, she was suffering from hypothermia. We needed a place out of the wind to warm her. I spotted an old, abandoned chicken coop, and Mr. Hall agreed. He carried the limp youngster as I led the mare, who followed her baby while voicing concerned grumblings. The door into the coop was short and narrow, and the ceiling of the shed was low — an entrance intimidating to most horses — but the mare entered without balking, following her offspring like a good mother, caring more about the babe than her own safety. Inside, she stood quietly watching over everything we did.

If love means caring about another's well-being more than our own, then postpartum mares demonstrate the definition of love. Mares can be dangerously protective, but when their babies are ill, they seem to understand that I am there to help. I remember a large bay quarter horse dam who hovered over her re-cumbent foal for several days. She watched attentively as I treated her colt for abdominal pain. On the third day, exhausted, she collapsed to her knees next to me with a grunt. I looked into her eyes and said, "Please lie down and rest. I'll take care of him." She rolled to her side and slept deeply, until her foal suddenly recovered, stood up, staggered over to her heaving side, and struck her abdo-men and udder with his front hooves, demanding she rise so he could nurse. She did, and they both lived happily ever after.

Rarely, a mare will refuse her foal, just as some human mothers have trouble bonding with their newborns. That is another emergency — the foal must get the first milk, the colostrum, within the first twelve hours of life to be protected from infection. Usually sedation and restraint of a reluctant mare are enough to get the two to bond. Once the neonate nurses, the hormone oxytocin flows through the mother, and she becomes attached, which is also true for humans. We have the same hormones (oxytocin, estrogen, progesterone, testosterone, and so on), and the same emotional responses to their effects.

Mr. Hall's mare showed a strong attachment to her filly, although her en-gorged udder indicated that the foal had not nursed for some time. I knew the neonate was dehydrated on the inside even though she was wet on the outside. Without the precious elixir of the mother's milk, hypoglycemia would be an-other problem. We found a 100-watt lightbulb in the chicken coop to provide heat, and we dried the neonate with towels. I fitted an old down vest on her body with the snaps connecting along her back, and I placed a catheter in her jugular vein to administer warmed fluids. I added antibiotics to the solution of dextrose and electrolytes because foals have very little immune protection and get infec-tions easily. Once the fluids had run and her body heat returned, the neonate began to come to life. She stood with our help, but her head hung down to the ground, eyes closed, as she wobbled, her spindly legs sprawled wide apart. She needed to drink her mother's milk to survive, but I learned long ago not to push foals to nurse; they just push back, and she might fall. Still, I gently nudged her to the mare, whose udder started dripping milk as she nickered to her baby and nuzzled her bottom. The foal replied with a weak, high-pitched whinny and a tail swish. She moved closer, head hanging below the udder as milk poured from the teats and dribbled over the filly's drooped ears. She shook her head, wrinkled her brow, and turned her lips down as if she felt annoyed. "Just open your mouth," I begged. After several long minutes, her whisker-covered muzzle opened, her pink tongue curled and reached up; she suckled without making contact at first, and then, to our relief, finally found the glorious goodness of her mother's nipple. The pair emanated love. I wondered how anyone could say animals do not experience love.

Although most scientists deny there is any evidence that animals feel love, there is a test to determine who loves you more — your dog or your spouse. Lock each one in the trunk of a car for an hour and find out which one is happy to see you when you let them out. This is a joke, but it rings true.

For a veterinarian like me, who observes animals in intimate situations, it feels intuitively obvious that animals love their offspring.

As I drove away, I thought of the many neonates I had saved, and in the warmth of fulfillment, I forgot about my pain and busy schedule. Then I remembered the precious ones that had died in my hands. My eyes blurred, my throat tightened, and the pain returned.

Veterinarians dance with death daily. When the phone rings outside of normal business hours, we get out of bed and go, not for the reward of money — other professionals with our education level earn much more than we do. Rather, we attend emergencies because we love the beautiful creatures and the ugly ones, too. We want to help them all. Their spirits touch us and bring us joy. At the same time, we have to make peace with tremendous suffering and our failures.

Ambulatory veterinary medicine is far from glamorous. Rather, it is blood in the mud, life-and-death decisions made outside of normal business hours, often during bad weather. Even in a hospital, during the day, emergency work is challenging. I faced this gruesome truth immediately after earning my doctor of veterinary medicine degree. In 1985, I began working at Animas Animal Hospital, where late-night and weekend emergencies were common and stressful, and I started my own equine ambulatory veterinary practice in 1988, which made me the sole person responsible for my client's animals around the clock. I loved my profession, but I whipped and spurred myself to attend emergencies past the point of exhaustion, and I agonized over the suffering of each animal I attended. My body felt like a sagging ridgepole about to splinter apart in the middle. After years of castrating untrained colts, watching horses thrash in pain, filing horse's teeth, treating bloody, maggot-infested wounds and pus-filled uteruses, and performing too much euthanasia, I needed a mental diversion to help avoid spontaneous combustion from burnout. At the same time, my close relations with animals from birth to death made me wonder how people could make certain statements about animals, such as "They do not feel love," "They are not conscious," and "They do not have souls." None of these made sense to me. I reached a turning point when I decided to research the world's religious, scientific, and spiritual teachings about the nonphysical aspect of animals. The quest to understand their spiritual nature became my passion and salvation.

Animas

I set out to understand the spiritual nature of animals, and in so doing, I discovered my own. Creatures great and small dragged me down a rabbit hole and through sacred tunnels into a world of dragons, shamans, gurus, lamas, monks, nuns, demons, priests, rabbis, preachers, scientists, clairvoyants, channels, mystics, animal communicators, and spiritual teachers. Those adepts schooled me and gave me refuge from the drama and trauma overburdening me. They introduced me to the *anima* — what Jungian psychology refers to as the animating principal present in all living beings.

Anima is the Latin root of the word *animal*. It means soul, breath, and life. Veterinarians share a personal relationship with the anima; we watch it drain from a body only to meet it again as a newborn foal or pup. Yet veterinary education rarely mentions it. We learn detailed information about bones, blood, and the other physical components, but little is said of the nonphysical aspect — the animas of animals. I now believe it is the most important part.

Firemen do not enter burning buildings or ascend to the tops of tall trees to grab a hunk of meat known as a "cat." They rescue a beloved family member, a companion. The incorporeal light in an animal's eyes reaches into our hearts. It touches us more deeply than any physical thing. We humans have the capacity to connect with the spiritual nature of animals; it makes us happy.

I wanted to be an animal doctor before I knew the word *veterinarian*. As a young girl in Wisconsin, I remember attending a stallion showing. The handler enumerated the attributes of the handsome, gray Arabian stud as I stared at a gray-haired man in the audience. He appeared humble and placid and wore a vest monogrammed with the veterinary emblem. I learned that he was the local large-animal veterinarian. Although I did not know him, something about his wise, yet nonjudgmental demeanor attracted me, and I longed to be like him. I had yet to learn how the fires of veterinary practice would burn and melt me before forging me into the person I aspired to become.

According to the American Veterinary Medical Association (AVMA), the suicide rate for veterinarians is four times higher than the national average. This may be due in part to the gory wounds and difficult procedures, the death and euthanasia, and the stress of long hours treating emergencies, but it may also be because of something I call "compassion overwhelm." We care too much. We tend to be compulsive overachievers who sacrifice our lives for the job, as did the cattle practitioner I read about once in the AVMA obituaries who drowned trying to save a calf stuck in a muddy pond. Or perhaps suicide appeals to us because of our familiarity with death.

No matter the reasons, my experience convinces me that an understanding of the spiritual nature of animals benefits the mental and emotional health of veterinarians and animal-loving people who anguish over the suffering of pets and wildlife. Furthermore, veterinary clients may harbor strong religious beliefs that influence their decisions, and we must show them respect and speak to them with wisdom. I hope the insights shared here provide comfort to those who live with, tend to, and love animals. Perhaps, once they learn of their beneficial qualities, some may even come to see and appreciate the simple beauty of maggots and other parts of nature we often abhor.

My goal, then, is to explore the world's religious, spiritual, philosophical, and scientific teachings about the nonphysical makeup of animals for the highest good of animal care, the human-animal bond, and the well-being of all concerned.

Each chapter explores a different religious belief system and offers three main approaches to the material. First, each chapter begins with a description of that belief system. In order for the reader to fully comprehend the tenets of Hinduism, Judaism, shamanism, and so on, the vocabulary must first be defined. Therefore, each chapter includes some history and definitions followed by an investigation into the religion's beliefs regarding animals. Then, throughout, I provide stories from my veterinary practice, offering further illustration of the concepts for contemplation. The third element explores the unfolding of my own spiritual growth — a concept I learned in the process — and how it changed me.

The struggle inside me first started in the 1990s and early 2000s when I practiced large- and small-animal ambulatory medicine out of a pickup truck in a rural mountain community in southwestern Colorado. Horses, llamas, alpacas, dogs, and cats were my primary patients, along with other wild and domestic flying, swimming, and crawling creatures. I drove day and night to attend to animals in beautiful places around Durango, Colorado, where the Animas River carves the landscape.[1]

The majority of my time spent researching spiritual teachings took place on the job in a pickup truck where the only way to learn was from audiotapes. Time off included further seeking by reading, praying, attending church services, meditating, chanting, going on a vision quest, attending retreats with a Tibetan lama and Buddhist nuns, questioning psychics and shamans, pursuing an animal communication apprenticeship, and conducting interviews with experts in numerous fields of spirituality.

During this period, I drove an average of a hundred miles or more a day. Fortunately, the scenery made that part of the job a pleasure, although the dirt routes on winding, mountainous terrain were often treacherous. The views of snow-covered peaks and enchanted valleys, tall ponderosa forests, and aspen groves soothed my stressed mind. Guardian angels held the wheel as I watched a bald eagle circling above or stared at an osprey sitting on a snag next to the Animas River. The occasional coyote made me giggle as he scooted across a field, looking over his shoulder as though trouble were on his tail. A bobcat, a bachelor band of elk, or a family of deer crossing the road at times caused me to hit the brakes. Drives to ranches were adventures into fantasylands of hidden canyons and mysterious ravines where cell phone service was lost. A tourist once told me, "You live in a postcard." To which I replied, "And I drive each day in the mud, dust, and snow that keep everyone else from living here."

My home is in La Plata County, a community full of colorful characters: ranchers whose families have lived here for generations, Native Americans, mountain hermits, Hispanics, and nonlocal people who moved to the country from cities. The stories I share about these people are true; at least, this is the truth as I remember it. Only the names of people, animals, and some places have been changed to protect privacy. One group of folks who live here, not found in cities, are cattlemen who breed livestock in small family-run operations. Their relationship with the animals they raise for income is often misunderstood. Humans and beasts depend on one another; one does not survive without the other. The rural life itself is foreign to many. The jobs of the mobile large-animal doctor intertwines with the country ranch life, and both involve hard physical work caring for beasts that can be dangerous. The veterinary work-force has evolved to focus more on pets in cities, and a shortage of rural veterinarians creates challenges for folks living in the country.

Today, in 2017, I spend most of my time practicing traditional Chinese veterinary medicine (TCVM) from my home office using acupuncture and Chinese herbal medicines. I now drive a VW Toureg when I attend to horses and llamas. I gave up emergency work in 2010, but I still end up treating an occasional wreck for a neighbor, friend, or desperate person who can find no one else to help. The evolution of my practice occurred gradually, but in 1995, it was aided by my interactions with two women who unknowingly set me on my course of discovery and inspired me to study the spiritual nature of animals — a young professional and an elderly local rancher, a Buddhist and a Baptist with contrasting views.

The Buddhist and the Baptist

"I'm a Catholic Buddhist" was the first thing Margaret told me. "The two are not mutually exclusive. Catholicism is a religion, and Buddhism is a practice, a philosophy."

In November, Margaret called because her elderly dog, Jaws, a vicious biter, was deathly ill and unable to stand. The end was near, but Margaret's Buddhist practice prevented her from electing euthanasia. Her teacher, the rinpoche, had told her not to kill. As a Buddhist, she had vowed to refrain from taking the life of any living creature. Furthermore, bad Buddhist monks reincarnated as dogs. Since Jaws could have been a human in a former life, or could become human in a later one, it was better for him to suffer his karma in this life so that he could have a better life in his next incarnation.

This philosophy was new to me, but I agreed to see the dog at Margaret's home and try to alleviate his struggle and help him pass. The reason Margaret called me was because she had heard that I used acupuncture to treat animals, and she hoped I might be more sympathetic to her dilemma. She did not want her pet to suffer.

Margaret and her daughter lived thirty-five miles away in a small mountain village. Snow-packed roads forced me to drive slowly to a little cabin in the woods.

I entered the house to find a dismal scene. The old terrier lay sprawled out in the middle of the floor, penned in by boxes and furniture. His coat reeked from urine and feces, since he soiled himself, being unable to stand. He looked miserable to me. The only cheery thing in the room was Margaret's smile. It was ever-present, like the serene smiles of Buddhist monks. Her gray eyes glimmered with a sense of peace and joy rather than the stressed-out glare most people express when facing the death of their pets.

"Would you like a turkey sandwich?" asked her plump daughter. The odor of urine dampened my appetite, and I declined as I tried to piece together the dog's medical history. With each question I asked, the two women responded with long stories of irrelevant information about their family history. Margaret's smile persisted in spite of the tales of her cruel father, a bitter man whom Margaret cared for up until his death, a situation now repeated with Jaws, who had been her father's dog. Margaret described her devotion to her father as a way to cleanse her karma.

"He was horrible to you, Mother. I feed Jaws turkey; is that okay?" the daughter asked me. "I have a thyroid condition," she added, patting her thick neck. "Are you sure you don't want a sandwich?" She pulled a tray with an entire turkey on it from the refrigerator.

"No, thank you," I replied, gratefully examining a copy of Jaws's blood work. From this, I understood that he had Addison's disease, or hypoadrenocorticism (a lack of cortisol and other hormones from the adrenal gland), which was a sequela to Cushing's disease, or hyperadrenocorticism (an excess secretion of hormones from the adrenal gland). He was being kept alive with the steroid medication prednisone.

Since this was my first attempt at treating a dying Buddhist land shark, I proceeded with caution. I carefully slipped a muzzle over his nose. "You're okay," I said to calm him. "Good boy." His eyes watched mine with uncertainty as I collected blood and performed acupuncture, but he never tried to bite, which told me he did not feel well enough to make the effort. I had just completed acupuncture training with the International Veterinary Acupuncture Society (IVAS), and I knew that properly placed needles would balance hormones, relieve pain, and improve circulation of blood and energy. I hoped it would provide some comfort to the dog until I had time to analyze the blood chemistry values. I left Margaret's house with two vials of blood and a turkey sandwich.

It took several days to grasp the meaning of the relationship Margaret had with her dog, the religious implications of the final days, and how to best help them both. At last, we decided to stop giving the prednisone to Jaws, hoping he would pass quickly and gently, which he did overnight.

The telephone woke me, as usual, and I wasn't surprised by the request. It was still November, the time of year when ranchers kill their geriatric horses.

A raspy voice said, "Doc, this is Polly Parsons. I have an old horse I need you to put down."

"Is he old?" I asked, not quite awake.

"He's about twenty."

I knew a lot of horses more elderly than that, so I asked, "Is he sick?"

"No, he's not too bad. I just don't believe in letting him get old and suffer."

Polly was a gray-haired rancher, a born-again Baptist, and in her mind this was the most humane thing to do. Ranch horses do not have warm barns with heated water tanks; they live like wild animals, out in nature, and if nature takes its course in winter, those "long in the teeth" get skinny, weak, and die. Ranchers like Polly preferred to "put down" their horses rather than leave them on pasture to weather the snow and bitter cold.

Each year, cowboy poets gather in Durango to spin yarns and narrate poems. I've heard more than one express the idea of killing old horses so they don't have to suffer. The story line of such a poem goes something like this: The cowboy goes out to shoot his old friend Buck. He aims his gun and looks down the sight into his friend's eyes, and then he remembers the times when the horse was his only friend out on the trail. The horse saved the cowboy's life more than once by protecting him from a cliff, a mean bull, or a deep bog. The two had covered a lot of territory together, and now the time had come to let him go. Then the cowboy notices that Buck doesn't really look that old. Maybe he has his birth date wrong; he's in good flesh, and mares still like him, too. Why, maybe he'll make it through the winter all right. The almanac says it might be an easy one. Talking himself out of the difficult task, the cowboy puts away his gun and drives on down the road.

Cowboys act tough and they sometimes talk rough, but they love their horses.

Occasionally, like Polly, a cowboy has called me to do the job. Once, a man hired me to euthanize a six-month-old, rye-nosed filly because she had trouble breathing. When I arrived, his older brother wanted to know why I was there. He asked his younger brother about the foal, "Why don't you just shoot her?"

"I don't want to shoot her. Do you want to shoot her?" he asked gruffly.

"I don't want to shoot her," said the older brother.

"Okay then."

Enough said, it was decided. I did the dirty work by lethal injection as the younger cowboy held back his tears, saying, "I should have put her down at birth, but I just couldn't."

I could tell Polly loved this black quarter horse, too, as she started reminiscing. "He was my husband's roping horse. People offered us up to ten thousand dollars for him; he was the best roping horse in the county. My husband died in my arms two winters ago. He didn't feel well one night, so I held him, and he just passed away."

Polly and I climbed the hill to the place she chose to bury the horse. Though almost twice my age, Polly surprised me with how briskly she could hike that rocky pasture. The horse surprised me with how quickly he died. He fell off the needle just as I completed the injection of euthanasia solution.

"Now he and your husband are together," I said.

"No, they're not!" shrieked Polly. "My husband's in heaven and that horse is just dead!"

"Well, what happened to the energy that was just there?" I asked, pointing to the body.

"I don't know. It's just gone," she said. And she started quoting Bible verses.

I know my way around a Bible, being raised a Christian, and although I've never seen a Bible passage that says animals do not have souls, I knew Polly believed they did not. I gave Polly a ride back to her little ranch house, while we discussed suffering and death and the laws of energy. She seemed to have all the answers according to Baptist teachings, until I asked, "Do you believe in reincarnation?"

"No, that's when you come back as an animal," she said.

"No, it's not. Well, not necessarily."

Then she quoted the Bible again. "There will be a new heaven, a new earth, and a New Jerusalem, and body and soul shall be reunited."

"That sounds like a description of reincarnation," I said.

"Well, maybe it is," Polly said with conviction.

Six months later, I tried to find the verse Polly quoted about the body and soul being reunited, but could not. I wanted to ask Polly, but I learned she had died. A friend suggested I speak to Polly's son, John. So I invited him to have tea with me.

John said the verse came from the book of Isaiah. He also told me about Polly's painful back injury. Surgery had not helped, and the pain medication made her sick. Then one day, no one could find her. John called out telepathically to his mother, asking her to tell him where she was. He walked across their many acres and found her, perhaps somewhere near where their old horse died. She had shot herself.

John also remembered the horse fondly; only he and his father could ride him. The steed had indeed saved his father once by carrying him safely out of a deep bog. He may have had some chronic foot pain, so maybe Polly ended his suffering, or maybe she was getting things in order before she killed herself. Or maybe there is more to the story than I know. For me, it was all as clear as the mud and dust that covered my truck.

I wondered how Polly had come to terms with the Christian teaching that suicide is a sin. I supposed that her belief about not letting horses get old and suffer also applied to her. This is where church dogma and her personal beliefs parted ways. Somehow, she must have reconciled the issue with God.

These two women, one a Buddhist and one a Baptist, made me wonder about the spirits of animals. Somewhere between the Buddhist notion that animals can reincarnate as humans and the Baptist belief that a dead animal is just gone, there must be the truth...and I aimed to find it.

Judgment and Pain

Remove judgment and pain disappears.

— THICH NHAT HANH

People tend to frown or laugh whenever they hear about Margaret the Buddhist and Polly the Baptist, and they judge the women as either cruel or crazy. The topic of religious beliefs stirs up strong opinions, childhood wounds, and deep emotions. Wars over belief kill millions. People cling to their faith, believing theirs is right and other faiths are wrong.

Similarly, numerous people have at times been upset with my interest in certain religious teachings. So, it was with trepidation that I began to write this book. I initially stood somewhere between Margaret and Polly, a Christian who believed in reincarnation. I had no idea how many other ideas existed. After much contemplation, I hoped that at the highest level of each teaching, a truth common to all existed. Hence, I searched for common elements among spiritual beliefs and found many positive, unifying themes. For example, the golden rule, "Love your neighbor as yourself," appears in the texts of African beliefs, Persian teachings, Buddhism, pagan practices, Christianity, Confucianism, Hinduism, Islam, Jainism, Judaism, Native American stories, Sikhism, Taoism, and Zoroastrianism.

Each religion or philosophical belief has an ultimate ideal or supreme spiritual power for which there are many names: God, Goddess, Truth, Yahweh, Allah, the Universal Life Force, the Lord, Brahman, Absolute Bodhicitta, Christ consciousness, the Tao, the Higgs boson, Source Energy, the Great Spirit, Buddha-nature, Adonai, the Force, DNA, All That Is, *Wakan-Tanka*, the Inscrutable, the Divine. As different as these names seem at first, they have a lot in common. Many teachings, including some Christian and Buddhist teachings, state that the supreme power resides inside us. Jesus said, "The kingdom of God is within you" (Luke 17:21), and each Buddhist has his or her own inner Buddha-nature. Because the divine is within us, we are encouraged to look there for the truth. The Buddha said, "You should trust the truth that is within you."[2]

In this book, when I refer to the "Truth" with a capital *T*, I mean the ultimate in spiritual teachings. The other kind of "truth" is a slippery, shifty, ever-changing viewpoint from a particular perspective. Ask ten people what they saw at the scene of a crime, and you'll hear ten different answers, all true.

In fact, if multiple stories of a crime are identical, the police consider them to be contrived. We each see things differently, and our story changes with time.

I originally intended to report the Truth about the spiritual nature of animals, but this became more challenging as a myriad of opinions surfaced for each religious teaching. Wide variation of beliefs exists even within a religion, sect, denomination, or church. With no agreement, no simple statements can be made about Catholic, Hindu, pagan, scientific, or psychic beliefs.

Dilemmas also mounted as I encountered my own negative judgments regarding spiritual teachings new to me. In order to present a fair and impartial view, I had to open my mind to new ideas, and the project quickly became entertaining.

Letting go of restrictive opinions opened my awareness to how judgment causes pain. During this time, my stomach hurt when I worried about my patients. My lower back ached as I anxiously raced to make appointments and get to emergencies quickly. I suffered in empathy for my clients and their animals as I wrote sympathy cards, agonizing over each death. Eventually, my mental, physical, and emotional health improved because of what I learned and share in this book.

The physical strain of ambulatory equine practice — floating horse's teeth, carving out sole abscesses, and castrating unruly two-year-old colts — took a toll on my body. Exhaustion followed long hours. I often came home late from an emergency too tired to cook. Dinner consisted of a glass of sauvignon blanc and dark chocolate. My dreams were shattered by images of steaming bowls of pus and memories of desperate, dying eyes looking up at me. A recurring nightmare plagued me in which my truck would not stop. My foot pressed the brake to the floor and the truck kept on rolling, always in different scenes, through intersections or down steep muddy roads with sharp turns. I was always unable to stop it.

Some puritanical work ethic drove me; I whipped myself like a self-flagellating penitent. I related to Martin Luther, founder of the Lutheran religion, whose life was discussed on audiotapes I played in my truck. The young Luther thought we had to work hard and suffer because we are evil, worthless sinners unworthy of forgiveness. My Lutheran upbringing and German heritage encouraged hard work. The fact that I was doing what used to be a man's job also put added pressure on me to perform and prove my value.

In January 1995, I visited a psychologist, who told me she could not help me; she said I had to change my life. This was easy for her to say but hard for me to accomplish. I was heavily invested in my job. I loved veterinary medicine — I still do — and I was good at equine ambulatory work. I had no idea how to change. She suggested I take days off, but this was no help. When I did, several people became furious with me when I was unavailable for their emergencies. One woman hated me because I took a day off and her horse died. Another veterinarian attended that emergency, but in her mind, I was to blame.

That kind of judgment hurt, but my self-judgment was even more painful. This was emotional pain. On the outside, I looked like a strong, fit, confident, capable woman. On the inside, I ached. The irony was that I had begged for this job. I had wanted to be a veterinarian more than anything else in the world.

Other colleagues seemed to be struggling as well. One of my veterinary school classmates and friends killed herself by drinking euthanasia solution in her Diet Pepsi. Another died when he rolled his truck on a late-night emergency call. In the neighboring county, a horse doctor committed suicide by shooting himself.

A turning point came for me about ten years later, at the funeral of a veterinary friend who died of cancer in a nearby town. She and I had planned to work together so we could have more time off. At the funeral, her two young children cried, while one person after another stood to tell stories about how this woman had come out at midnight, or on Christmas Eve, or 10 PM on the fourth of July, or 5 AM on a Sunday morning with her children sleeping in the truck. My friend lived with the stress that a large-animal veterinarian works under day and night. She used all her energy to help other people and their animals, but this meant she didn't have enough left to stay healthy herself. It hit me like a bullet. I was doing the same thing, killing myself by working nonstop. The challenge was to find a way to change.

My animals helped me find relief. At night when I tossed in bed, my cat would come nuzzle my cheek and snuggle into my armpit. Her nonjudgmental nature touched my soul and gave me peace. One of the most beautiful characteristics of animals is their nonjudgmental, forgive-and-forget nature, which offers us relief from the pain of constant human appraisal and self-loathing.

Opinions about right or wrong, good or evil, represent a dualistic belief system that involves moral judgment. Some animal behavioral scientists and religious scholars believe that animals do not develop moral opinions. I question this supposition and explore it more fully throughout this book. We have no way of knowing for certain what an animal thinks or feels. We have trouble trying to understand other humans. We do know that a dog learns human rules about right and wrong, but the animal may not share our human perspective. A dog will romp through the house with muddy paws wondering why their person is so upset, and moments later, the episode forgotten, the dog wants to play; tomorrow, the dog will have no remorse or qualms about spreading mud on the carpet again. At the same time, animals who live in social groups, such as wolves, appear to have rules that teach them right from wrong in the pack. They argue about these rules and they also forgive, which leads me to consider the idea that some level of morality exists among some animals.

Children, especially, come to appreciate an animal's characteristic nonjudgmental, forgiving nature. As a youth, I found solace with my Shetland pony, Earl. He played with me and accepted me even if, in some human's opinion, I was fat, stupid, ugly, or wrong. One man shared with me that he hid with his dog in the doghouse after hearing his father refer to him as "stupid."

Even the great philosopher Socrates appreciated this quality in a dog. As Thomas Cleary wrote, "Socrates used to take shelter in a barrel with a little dog. Some of his students asked, 'What are you doing with this dog?' Socrates said, 'The dog treats me better, since it protects me and doesn't annoy me, whereas you desert me and yet annoy me, too.'"[3]

When our pets accept us without judgment, it feels like love. Unconditional love means being accepted under any conditions, right or wrong, smart or stupid, Buddhist, Baptist, Muslim, or Hebrew. As I explored religious beliefs about animals, I considered the issue of moral judgment for several reasons: First, nonjudgment is another core Truth in the spiritual teachings of Jesus, the Buddha, and many other religious leaders. Jesus said: "Judge not, so that you may not be judged. For with the judgment you make you will be judged, and the measure you give will be the measure you get. Why do you see the speck in your neighbor's eye but do not notice the log in your own eye? You hypocrite, first take the log out of your own eye and then you will see clearly to take the speck out of your neighbor's eye" (Matthew 7:1–5). The Buddha told followers not to set themselves up as a judge of others or make assumptions about their motives. You can destroy yourself by holding judgments about others.[4] But also, I wanted to investigate what I could learn about whether animals differ from us in terms of a moral sensibility. Finally, spiritual teachers like Thich Nhat Hanh and Eckhart Tolle teach that pain results from judgment, and I needed to learn to free myself of negative judgments, self-judgment, shame, and pain.

According to Eckhart Tolle, "If you stop investing [the pain] with 'self-ness,' the mind loses its compulsive quality, which basically is the compulsion to judge, and so to resist what *is*, which creates conflict, drama, and new pain. In fact, the moment that judgment stops through acceptance of what *is*, you are free of mind. You have made room for love, for joy, for peace."[5]

This idea of accepting rather than judging what is arose dramatically for me when I met Dana Xavier. On a Sunday in early December 1999, Dana called with an emergency — her stallion had lacerated his face — amid a blizzard of wet snow. Dana lived up a steep hill, on a mesa, and her driveway was gravel and clay with no guard railings. It was a tense drive, even though the four-wheel-drive Ford handled the slimy mud. As I sutured the horse's wound, I asked the small woman what she did for a living.

She said, "I'm a clairvoyant."

"You mean a 'psychic'?"

She nodded. She had a curious way of communicating; she didn't offer much, just giggled a lot.

About a year later, in March 2001, I made an appointment with her to discuss the spiritual nature of animals. During our talk, I told her how badly I felt if a patient did not heal, and she laughed. Then she laughed some more. She laughed until she cried and almost fell off her chair.

I felt insulted. "I'm glad I can be a source of amusement for you."

"I'm sorry," she said, wiping her eyes and regaining her breath, "but it's not your fault if an animal doesn't get better."

"Really, well, then what do people hire me for?"

"Oh, you help, for sure, but there's a lot more going on than just what you do. Maybe the animal had other plans. Maybe the animal and their human are trying to learn something together."

"Animals have plans?"

"Yes."

"So, what am I supposed to do when an animal is not improving?"

"All you can do is pray for the animal and the people to receive the healing they want."

I had trouble with this and thought Dana was a bit crazy.

Then she said, "Don't judge the situation. Judgment is pain."

Clearly, I had much to learn: about the realm of clairvoyants, about the "plans" animals had, and about how animals and humans learn together, but that had to wait for future meetings.

How to Heal

I decided to put what Dana told me into practice — to stop judging myself, others, and situations. The perfect opportunity presented itself several months later with Dawn, a lovely woman who first hired me to treat her horses, then to perform acupuncture on her six-year-old shepherd-mix named Apache.

It started when Apache jumped out of Dawn's truck and ruptured a spinal disc, becoming paralyzed in the rear end. Dawn had rushed her to Albuquerque for spinal surgery, and she had called me from there the following day.

"The surgeon said I should think about putting Apache down because she has no feeling in her hind limbs. Karlene, what am I going to do? I can't lose this dog."

"Does the surgeon know you were once paralyzed?" I asked.

"No, I didn't tell him."

"And look at you! You're walking around just fine. And you were paralyzed from the neck down."

"Yeah, they told me that I'd never walk again."

"Right, so we won't listen to him," I said.

We started electro-acupuncture on Apache right away — twice a week at first and then weekly — and after five weeks, post-op, Apache had regained feeling and movement from head to toe. She wagged her tail and had bladder and bowel control.

One day about this time, as I drove up to Dawn's ranch to treat Apache, I saw her waiting for me next to the barn, a tall, slender redhead wearing a long, russet denim jacket and irrigation boots, looking like a model for a farm-supply catalog. She also wore a frown and was holding a lead rope connected to her mule, Candy. I knew something was wrong. It is common to have a client say something like, "While you're here, doc, can you look at...," and I will add another animal to the day's schedule.

Dawn was worried that Candy might have "stringhalt," a jerking motion of the hind limb caused by nerve damage. I started my exam by making friends with the mule and trying to lighten Dawn's mood.

"Howdy, Candy girl," I said as I rubbed the mule's face. "You're a good girl." She rubbed her head against my hand. "You like me, don't you?" I cleaned the sleep from her eyes.

"Horses must like us," I told Dawn. "Otherwise, why would they put up with us? Some people think they only love us because we feed them. Well, it may be 'cafeteria love,' but how does that differ from men? They always say the way to a man's heart is through his stomach. I remember a time when I was in the Denver airport. A man on my flight kept staring at me with a hungry look, practically drooling. We boarded the small plane, and once in the air, he unbuckled his seat belt and came back to talk to me. He bent toward me and said, with a southern accent, 'I bet you make great biscuits and gravy.' I said, 'I've never made biscuits and gravy.' That was enough for him. He turned around, went back to his seat, and never looked at me again."

Dawn curled the corners of her lips and rolled her eyes, not amused by my story.

"I reckon men are no different than any other animal. At least you guys know what you want," I told the mule. "You like food and you like your face rubbed. Let's see how you like it when I palpate your stifle joint."

I slowly felt my way down Candy's back to her hind limb, palpating the kneecap area. "Walk her around a little. See that clicking motion by the stifle? That's not stringhalt. Stringhalt is when the limb jerks up toward the belly. This is upward fixation of the patella — the ligament of the kneecap is catching on her femur. It happens in horses with a straight-legged confirmation. You just need to exercise her to strengthen the muscles to hold the kneecap in place. Take her riding in the sand wash; that will tone her up."

"I can't go riding. I have to take care of Apache. She just panics if I get out of her sight."

This is the challenge I was not taught in veterinary school — how to manage people's lives. I had learned anatomy, microbiology, radiography, and surgery, but nothing about how to help people manage their problems. Dawn had good reason to be depressed with multiple ailing animals to care for. I really wanted to help. "How is Apache?"

"She's okay. Come see her." Dawn carried Apache from the front seat of her Dodge Ram and stood her upright on the ground in front of me. "Look," she said, still not smiling. "She can stand."

"Oh, that's great!" Apache collapsed, and Dawn lifted her onto my tailgate, where I had a rug spread for a treatment table. "Hi, Apache, you look so good." However, for the first time, I noticed a cowering expression in the dog's eyes as she looked at Dawn, prompting me to ask Dawn, "What's going on?"

"Oh, sometimes I think that she's never going to get any better than she is right now." Dawn sighed. She gave Apache a worried look, which made the dog cower even more.

Dawn was worn out, even though Apache had progressed significantly. "Dawn, you have to hold this dog in a vision of health," I urged. I paused to think of a way to explain. Then I related a story that shows how animals think.

"My friend Betty has a big shaggy malamute dog named Harry, and each summer she shaves his hair short. She worried about shaving him again this year because he always hides under the table for a week. I explained to her that Harry hides not because he's embarrassed by his hairdo. He acts embarrassed because everybody looks at him like he's a geek and laughs at him. So, I told Betty that the next time she trimmed Harry, she should tell him he's a stud. Well, that worked. She told him, 'You look so handsome,' and he strutted around the house proudly.

"During my years of practice, I have found that animals mirror us; they reflect our thoughts. How would you feel if every time people looked at you, it was with pity in their eyes, or if people told you that you were stupid every day? You would feel the way they think about you. You have to look at Apache like she is getting better and encourage her."

Dawn replied, "Well, I need encouragement, too. I can't do anything. I have to take her everywhere. My back gave out the other night when I picked her up. What am I going to do if she doesn't get better? I'm really worried."

I pointed out, "She has bladder control. That's huge. And she can stand!" Dawn nodded, and I continued, "It has only been five weeks. She's getting better every day. We can help her; it's time to do more physical therapy."

I wrapped my fingers around Apache's hind limbs to feel her femoral pulses. I looked at the color of her tongue and placed the appropriate needles in her back, hind limbs, and paws, then I connected the electro-acupuncture wires to send a current through the needles. Apache relaxed and enjoyed the attention as Dawn stroked her head.

I hoped to hear some good news and asked about Dawn's horse. "How's Poco doing since the horseshoeing clinic?"

"He's lame. He threw the shoe off his bad foot."

I had treated Poco off and on for about a year, and he still had an intermittent left front lameness and stood with that foot pointing out in front of the other. I had referred the horse to a lameness expert and horseshoer, who applied a special shoe that Poco promptly threw. The thought of my inability to diagnose and treat Poco's problem gave me a stomachache. Due to my anxiety over unhealed patients, and the worries of clients, I probably had an ulcer; my heart palpitated; I had chest pain and shortness of breath; and all my joints ached. One of my legs was shorter than the other, and my lower back was in spasms all the time from a protruding disc. I was a mess, and if what the clairvoyant, Dana Xavier, had taught me was true, I needed to heal myself by allowing animals and their humans to accept responsibility for themselves.

Pain educates; we learn best from hard lessons. Sometimes conditions do not heal, and as long as I do the best job I can and have the best intentions, a lack of healing is not my fault. Conversely, I cannot take credit for a cure. I do not heal my patients; healing happens from inside each being, just as skin cells grow to fill in a wound. I am not healing the wound; I am not in control. I do the best I can to clean and protect the wound. I nurture and support the process while trusting that the innate capacity to heal occurs.

That day, I began to accept that I am not responsible for healing. I do the best I can and know that people and animals are working together to learn and may have other plans. I started to let go of judgments about my inability to fix every ailment; higher forces are at work beyond my best intentions. I left praying that Dawn, Candy, Apache, and Poco would receive the healing they each wanted. At the end of Dawn's long, gravel driveway, a huge snake slithered out of the grass in front of my truck. I stopped and went over to check it out. A bull snake, about five feet in length, turned to face me. He curled with his mouth open and shook his tail like a rattlesnake, a common tactic of the bull snake. He knew that behavior scares things away. He made me look twice — no rattles. He was beautiful and fat, as big around as my arm. Then I saw why he was so defensive. Something had taken a bite out of his side, leaving a hole the size of a chicken egg. Flies hovered over it, the parents of the tiny maggots moving inside the wound.

Part of me wanted to catch him and clean his wound. Yet a deeper part of me knew better, and I heard a voice in my head say, *Let it be*. There was no redness or swelling, and the skin around the wound was shiny and smooth. He was a healthy fellow. The maggots were keeping infection away, and the clay packed in the lesion made the perfect bandage, like a flexible plaster cast. Nature was healing the bull snake just fine without my help. So, I let him be, knowing that healing is part of the intrinsic quality of life. Snakes entwine the staff in the medical and veterinary emblems, symbolizing healing and transformation. Perhaps this snake was a wounded symbol of healing, an omen intended for me as well as my patients.

Fear and Fortitude

A veterinarian faces fear every time he or she meets a vicious dog, a screaming cat, or an aggressive horse. Fear on the animal's part drives defensive behavior, so we make an effort to proceed gently to calm the animal. I will not win a fight with a horse. Somehow, we animal doctors have to stay calm and confident and convince our patients to cooperate. Animals sense people's energy. When horses sense fear from a person or other animal, they take advantage and push the other around, biting and such. I learned as a child to hide my fear from horses and act tough. It takes fortitude to keep fear at bay. Thankfully, the majority of animals allow veterinarians to treat them even when the procedures, such as injections, do not feel good. All horses, by virtue of their size, are dangerous. They can knock a person down, bite, kick, strike, and step on feet. A new horse patient often tests me. I have always been amazed at, and grateful for, the many horses that respond to simple verbal commands, like "Stand" or "Quit." They get the message that I am not intimidated by them, and they obey. The same is true for dogs. A stern, one-word command works like a charm. There have been times when I have been intimidated by uncooperative horses and something unpremeditated has taken over me. Some kind of fortitude sparks a focused energy I send through my eyes and voice so intensely that it convinces those horses to stop misbehaving, and they obey me. Every veterinarian in practice is familiar with the fortitude required to face unfamiliar, growling dogs, hissing cats, and rearing horses.

Fortitude exists in all beings, human and animal, and does not arise out of an intellectual process. It springs from spirit, the invisible force we feel in ourselves and others. It emerges most forcefully when we face danger or a threat. Ever try to restrain an angry cat? Watch out! A cat can channel the fortitude to make two Doberman pinschers back away without touching the feline. The spirit flows through the eyes, and the voice tells those dogs to "back off," and they do.

In other words, I've discovered firsthand that there is more to living beasts than fur and feathers and physical parts, and my journey to explore the spiritual nature of animals is in part an attempt to understand this important component of life.

I first discovered this spirit in myself through martial arts training. By the late 1980s, I had earned a red belt in karate, and I was looking for a new teacher, or sensei, to take me to the brown belt level. The man I hoped would train me was well respected by other karate students in town. He had a fourth-degree black belt in Shotokan karate and was a master of tai qi chuan. He had also been a mercenary soldier. I met him to ask if he would teach me, and he said he wanted to spar. I was not experienced in fighting, especially not against such a powerful man. He stood over six feet tall, with a shaved head and camouflage fatigues. Still, I agreed. Though we wore padded gloves, it still stung when he popped me in the forehead. The blow hurt my neck, which made me mad, and I charged at him with my fists flying. He crossed his arms in front of his face, laughing. "Whoa, whoa, whoa, okay, okay.... I'll teach you. You have brown belt spirit."

Spirit was the term my teacher used to describe the energy that moved me to go after someone I could never beat. That energy came from inside me without conscious choice. The same energy has exploded out of me at times when a horse behaves dangerously. In moments of intense fear, thought stops and something takes over me, just as it happened when my teacher hit me. I will look the horse in the eye and command, "You stand still!" in an eruption of intense, palpable energy that even frightens me — after the event. Amazingly, the majority of horses obey, stop rearing or pushing me, and I pet them and tell them, "You're okay." I understand that they are frightened, too, and they seem to bond with me as if I were their leader. My reactions feel like spirit that arises from some nonmental instinct rather than a learned skill, and it is directed through my eyes and voice, just like an angry cat.

Fear is our first opponent in a fight. It paralyzes us and must be overcome. In karate, I was taught to fight with "no mind." Thought is too slow; during a fight, there is no time to think about what to do, so the body is trained with hours of repetitive moves to learn how to react when attacked. Furthermore, when I stop thinking, fear dissolves. Whenever I sparred in a no-mind manner, I did not remember what happened. The only memory I had was of the first technique after the command to start and of the last technique before the order to stop. The men in class would tell me how my opponent had me cornered, yet I landed a spinning back fist followed by a back kick, and so on, none of which I could recall.

I also observed my body moving in ways I never learned. Somehow my body defended itself when I was attacked. On one occasion a girlfriend jumped on my back with her arms around my neck, saying, "Okay, karate girl, what are you going to do?" I thought, *Well, I'm not going to poke your eyes out; I'm not going to break your ribs with an elbow strike.* I didn't want to hurt her, so I gave up, and the moment I quit thinking, something took over. My right foot went back and my torso bent forward, quickly launching her over my head onto the grass. She looked as stunned as I felt. It all happened as if some force took over my body.

My sensei also called that force or spirit "qi" (pronounced "chee"). In karate class, we learned how to focus and direct qi for more powerful movement. Qi gives even small people, and cats, great strength.

Sensei said, "You can stop a fight with a look." This is exactly what cats do, and it has been proven true for me on several occasions. Large men have backed away from me when spirit was directed through my glare. But again, I never intentionally think to do this. Fortitude just rises up out of me when danger enters my space. In short, something animates me that is not related to my conscious decisions.

Sensei's spirit radiated at least twenty yards whenever he demonstrated a kata, a patterned series of karate moves. His face appeared serene. He moved smoothly as if he was swimming, yet I found myself backing away. The power that radiated from him was not physical; it was energetic spirit. The same spirit or power emanates from animals, especially in the wild.

One day, while cross-country skiing with two friends in Rocky Mountain National Park, we spotted a large herd of elk bedded down about a hundred yards away. The cows were startled and struggled to move off in the deep snow. Then a large bull stood and glared at us. The force of his gaze knocked me backward into the snow. He radiated his energy so impressively from such a distance that it physically struck me. My friends and I agreed to go another direction so we would not disturb the herd. From karate and this bull, I realized that nonphysical energy emanates from human and animal bodies.

Through karate training I learned to appreciate Eastern philosophies and the concept of how to move energy. I understood how to direct qi, and I could feel it from others. About that time, I felt burned out with my job. Animals acted like they hated me. Everything I did seemed mean: castrating colts, deworming, giving injections to foals that acted terrified, floating horse's teeth. The animals lacked an appreciation for these procedures even though I intended to help them. Even more difficult for me was my inability to cure so many problems. Sometimes people could not afford the best treatments; other times the drugs I used made things worse.

Then I learned about an acupoint on the inside of a horse's lip that changed everything. Pressure at this point causes horses to calm down, lower their heads, and lick their lips. (Licking tells me a horse likes what I am doing.) The stimulation of this point releases endorphins, the body's own opiates. I came to use it every day.

One day a man called me to give his horse annual vaccines, deworming, and float his teeth. When I arrived, the man informed me that the horse could be difficult, maybe a bit mean, and that he hated needles. He thought a woman might have a better chance of getting along with the horse.

I approached the gelding easily, petting him gently and stroking his soft muzzle. I pressed my left index finger in the center of his upper lip until he seemed comfortable with my touch, then I slipped it in between his upper lip and teeth and pressed on a depression between the two front teeth where the lip meets the gum. The horse leaned against my finger, lowered his head, and licked. Then I gently but firmly pinched his upper lip with my fingers, digging the fingernails into his lip and the acupoint in the middle of the upper lip. His head came up a bit at first and then he closed his eyes, head lowered again, and licked more. I had a twitch (a foot-long metal pincher that looks like a nut cracker) on my left forearm and slowly slid it up to pinch the lip as I released my fingers. The man squeezed and wiggled the twitch as I easily administered intravenous sedation. I completed all the procedures without any fighting. I was happy to accomplish the job and stay safe. The twitch, used properly, is an excellent tool that stimulates acupoints causing the release of endorphins and other happy chemicals. It has kept me safe countless times.

This amazing acupoint made me want to learn more about acupuncture. Then I heard about the International Veterinary Acupuncture Society (IVAS), and I signed up to attend in 1994. There I discovered my natural aptitude for traditional Chinese veterinary medicine (TCVM). It resonated with me, and I felt at home with it. After an internship and an exam, I became certified as a veterinary acupuncturist in 1997. Then I started the IVAS herbology course and met Dr. Xie, a third-generation Chinese veterinarian who taught at the University of Florida veterinary school. Not only is Dr. Xie extremely knowledgeable, he is also one of the kindest people I have ever met. His students call him "Shen." In TCVM, *shen* is the spirit that resides in the heart. That fits Dr. Xie. For me, the man is a sage — a wise, humble master. Whenever a critic of acupuncture in the veterinary profession disputes the validity of veterinary acupuncture, rather than argue, Dr. Xie always advises us to "just make good qi." Dr. Xie came to Durango to treat patients with me on his next trip to the IVAS herbology course where we met. Then, in October 2000, he invited me and my husband to travel to China with him and a group of other veterinarians for an advanced TCVM course. He asked me to share a lecture at the conference about the wild horse foot, and how horses in the wild keep their hooves trimmed naturally and stay sound, in comparison to the domestic, shod horses that have so many hoof problems and foot pain. From there I finished my Chinese herbology training with Dr. Xie, at his school, the Chi Institute, and I also learned tui na manual therapy and food therapy.

I felt much better adding TCVM to my practice. About 85 percent of my patients enjoy acupuncture treatments; dogs like it because they get "cookies" (as treats) and "opium" (or the opiates that are naturally released by the acupuncture) — everybody likes cookies and opium. The body's pharmacy releases over 360 different chemicals, such as endorphin, encephalon, serotonin, bradykinin, antihistamine, corticosteroids, and so on, that relieve pain and calm anxiety. Acupuncture feels more nurturing to me, and the Chinese herbal formulas have helped many animals when drugs have not. I feel blessed to have studied with Dr. Xie, who taught me so much that improved my life and the lives of animals.

I am also blessed by the friendships I have at a local veterinary clinic, Animas Animal Hospital. From 1985 to 1988, I worked at this small-animal veterinary hospital in downtown Durango as a new graduate from veterinary school. One of my favorite things about Animas Animal Hospital is all the wildlife brought there for treatment. The doctors there work closely with the Division of Wildlife, and they treat injured wild animals for free. In addition, for many years, one of the veterinary technicians carried a wildlife rehabilitation license.

My relationship with Animas Animal Hospital continues, even though I have not been employed there for many years; I still have a key. I am grateful for my friendship with both the original and the current owners, which allows me the freedom to use the facility and its staff. Somehow, I am grandfathered in like an old fence that delineates a property line although the survey disagrees.

During the 1990s, five veterinarians worked at Animas Animal Hospital, and they were always busy with medical, surgical, and emergency cases. I often stopped by when I traveled through town. I used the microscope and X-ray processor, talked with the staff, picked up deliveries from veterinary suppliers, consulted with the other doctors on cases, and visited the "prisoners" in the back of the hospital, the area where the cages were located and treatments were done.

On one occasion, I entered through the back door to see a tiny orange kitten hanging from the bars of her kennel, calling, "Mew, mew, mew," just as Dr. Walter Truman walked into the room holding a peregrine falcon. The bird was calm, showing no fear. He just stared into my eyes like he found me as interesting as I found him. His spirit felt strong and his feathers were gorgeous.

"Look at those feathers," I said.

"They're in good shape," said Walt. "He hasn't been on the ground long; he was hit by a car. Look at how blue the sere is." I studied the area around the nostrils as Walt continued. "He's young, probably one or two. This is the third peregrine I've seen hit by a car on the highway by Yellow Mesa. It feels like the wing is broken. Stick around; I'm about to X-ray him."

"You're holding the world's fastest creature in your arms," I said.

"Yeah, they're amazing. They've been clocked at over two hundred miles an hour. I've seen them climb up into the sky over a duck pond so high you can't see them, and then, when the ducks are flushed off the pond, dive down and take a duck's head off." Walt walked into radiography as I played with the kitten's paws that reached out through the cage bars to touch me.

I glanced around the room to see what other interesting cases were in for the day. "Whoa, is that a fox?"

"Isn't she sweet?" answered Dr. Jane Becker. "A couple brought her in. They swerved their car to miss her mother and bumped into her."

Tears formed in my eyes at the sight of the beautiful fox kit. Her pointed, black nose and silver fur captivated my attention. Her deep, dark eyes looked into mine with complete calm. She seemed peaceful, reminding me of stories about the Galapagos Islands, where the first visitors found that wildlife had no fear of humans. A person could walk right up to an animal and pick it up. Only after humans started taking specimens and doing research did they become afraid. After that, the young learned from their mothers to run away. "This girl obviously doesn't fear us yet," I said.

"She is innocent. Of course, we aren't handling her. We want her to stay wild. Fortunately, there's nothing wrong with her. Unfortunately, Walt says she's too young to know how to feed herself. I tried to tell the people to leave her where she was when they called, but they were already on the way in with her."

People often find wildlife and think they need to rescue the creatures, but the reality is wildlife may do better without human intervention. A fawn, for example, is often left by the doe in a place to remain still until she returns from foraging. People find the fawn and take it home, thinking it has been abandoned, while the mother only went shopping and will return to find her baby kidnapped.

I felt so bad for the fox and her kit being separated that I almost cried. Without a mother, how would she learn to hunt? The people who captured her meant well, but with wildlife, we are better off letting them be. We have to remember that nature's way is best.

I walked back to the X-ray table to check on the picture of the peregrine as Dr. Truman was reading it. "Oh, that's too bad," he said. "Both the radius and ulna are fractured. Usually if one is broken, the other bone acts as a splint, and the wing heals really well. But with both broken, I'll have to do a surgical repair with a Kirschner external fixation apparatus. Then I'll send him to the rehab center over in North Fork. They have a flight cage there the size of three basketball courts."

The spirits in the falcon and the fox touched me. I felt some inner presence streaming from their eyes, some unnamed light flowing through them. The force from the elk that knocked me down and the force that moved my body to fight felt powerful even though they were invisible. This was the beginning of my investigation into the spiritual nature of animals.

CHAPTER 2

The Beginning:
Creation and the Garden Paradise

In the beginning, that is, in mythical times, man lived at peace with the animals
and understood their speech. It was not until after a primordial catastrophe,
comparable to the "Fall" of Biblical tradition, that man became what he is today —
mortal, sexed, obliged to work to feed himself, and at enmity with the animals.
— Mircea Eliade[1]

In countless cultures, ancient stories tell of a golden age, in the beginning of
time, when people lived in peace with one another and with the animals in a
garden paradise that provided abundant vegetation as food for all. The animals
communicated with the people, and they understood one another. Then something happened that changed everything.

Even though the stories of humans and animals communicating seem like
fables, their ubiquitous presence lends credence to the possibility that, at one
time, people did indeed believe understanding was shared. Although many consider paradise myths fiction, they also have historical bases and may reflect on-
going truths about our world and our relationship with animals.

Richard Heinberg, in his book *Memories and Visions of Paradise*, concludes
that the two — historical fact and symbolic metaphor — are intertwined. History, as a discipline, originated in myth; both are stories of our past. History
exists in myth as surely as myth persists in history. Myth functions through symbolic expression belonging to the realm of the mystical rather than to that of

reason.² Myths are ways to convey universal truths, and they serve to connect the visible and the invisible, earth and heaven.³ Mythical stories invoke images we understand on a psycho-spiritual level about a nonphysical existence we lack words to describe.

Having said that, I want to make a disclaimer that applies to all the stories, religious teachings, and myths in this book. I examine these stories looking for the shared, universal Truths they contain. In fact, I find over and over again that the main difference among the world's spiritual teachings is the vocabulary. We fear and fight each other over semantics. I too had fear of the new teachings I encountered during my investigation for this book until I opened my heart to understanding them. Then I discovered such beauty and love in the countless similarities between myth and science and religious teachings. Focusing on the common Truth is joyful and fascinating, whereas when we focus on the differences, we find reasons for fear and hate. I have no interest in convincing anyone of any particular religious belief, nor am I evaluating different myths to decide which are true and which are not. I want to explore all these stories, both to enjoy them and to learn what we can from them. To put this another way, I suggest that all these myths and teachings may share some universal Truths and may even reflect some accurate historical information, and yet none may be literally true in every respect, and I simply care not whether anyone believes any of it. There is enough drama in my job, and the study of spiritual teachings about animals has provided relief and entertainment for me. I hope you will join me in that spirit.

Once we open ourselves to the language of different stories, we understand why mythologist Joseph Campbell says that all myths are true; the psychic unity of humanity is expressed through mythology. He explains in his video *The Power of Myth* that myth brings us into the consciousness of the spiritual — precisely where I choose to explore. Creation stories are simply metaphorical descriptions of the indescribable. They are both metaphorical and true for the people who tell them.

Many beautiful, interesting teachings come from the past, when people used symbols of their ancient cultures. We can look at these symbolic stories and understand the Truth they convey without abandoning our religions or adopting those of others. Furthermore, we gain understanding and lose fear of other beliefs, which is what happened for me. I feel quite comfortable now entering the religious buildings and ceremonies of many different groups to share in the celebration of divine love and life because I no longer fear the unknown.

I believe it is important to say this because I am often shocked by the number of people who express intense anger about spiritual teachings they do not share or agree with, including what seems to me to be a benign paradise myth. Strong opinions and emotions are common with any discussion of religious ideas. This is one reason I want to emphasize that I am not asking readers to believe any particular story. I invite everyone to decide for themselves what to believe.

Regarding paradise myths, some people dislike them because they are often patriarchal narratives that portray a male creator. Indeed, according to L. Robert Keck in *Sacred Quest*, the skill of writing emerged on the human scene about the same time as did patriarchy, so perhaps it is not surprising that our first recorded creation stories reflect a patriarchal worldview.[4] In this book, I use inclusive language as much as possible, but I also tell religious stories in the ways they were originally told, even when that includes male-biased language and conceptions.

Another concern people sometimes have with religious myths is that they spread what might be considered wrong, inappropriate, or even "evil" teachings. Again, my goal is not to judge whether certain spiritual beliefs are right or wrong, appropriate or inappropriate, but to examine them for what they might tell us about our spiritual beliefs about animals. I searched far and wide, seeking a broad range of geographic and cultural perspectives. What I offer are the results of my journey of exploration, and I invite readers to draw independent conclusions. I urge you to evaluate what you agree with for yourself and to scrap the rest. Creation stories and paradise myths are important to understand because they are the foundation for modern religious beliefs. Human spiritual practices today build upon our shared ancient history.

And so, in this chapter, I examine four ancient myths from four very different cultures and areas of the world: the Native American Hopi, the Australian Aborigines, the ancient Hebrews, and several African tribes.

These creation myths share the common theme of people living at peace with animals and communicating with them before being expelled from a garden paradise. There are five other themes to notice: 1) Sound is an important vibrational substance of creation; 2) human and animal forms are made from the earth and given names; 3) the spirit or power of the creator is within everything, including animals; 4) specific creatures usually play roles in the stories, such as the spider and the snake; and 5) creation is destroyed and animals are saved.

The Hopi

The Native American Hopi tribe lives in northern Arizona. The word *Hopi* means "peace." Here, I paraphrase the Hopi creation story found in *Book of the Hopi* by Frank Waters.[5]

As the story goes, in the beginning, there was only an immeasurable void that lived in the mind of Taiowa, the Creator. Taiowa created So'tuknang, who became the creator of life. So'tuknang molded forms and made the universe half solid and half water.

This was the first of four worlds described in Hopi mythology, and each new world followed the destruction of the previous one. In the First World, So'tuknang created Spider Woman to be his helper. Spider Woman mixed earth with saliva and molded it into two beings. She sang the creation song over them, and they came to life. "All sound echoes the Creator," said Spider Woman, and the world was made an instrument of sound. She sent these twins to opposite poles of the earth to keep the world rotating properly. According to anthropologist Jeremy Narby, twin creator beings are another extremely common theme in world mythologies.

The earth vibrated with the energy of the Creator. From the earth, Spider Woman created trees, bushes, plants, flowers, all kinds of seed and nut bearers, giving each life and name. In the same manner, she created all kinds of birds and animals, and the power worked through them all.

Spider Woman made four male human beings, then four female partners, each of four different colors: yellow, red, white, and black. She granted them pristine wisdom, and they understood that the earth was a living entity like themselves. As Waters describes it, the bodies of humans and the earth were constructed in the same way, with vibratory centers along an axis, and this mirrors Hindu and Tibetan mysticism. The channeled material of Edgar Cayce also refers to God creating five colored humans at once.

The first people multiplied and were happy, although they were of different colors and spoke different languages. All humans, birds, and animals felt as one and understood one another without talking.

Waters writes, "They all suckled at the breast of their Mother Earth, who gave them her milk of grass, seeds, fruit and corn, and they all felt as one, people and animals."[6]

Then entered Lavai'hoya, the Talker; he came in the form of a bird called Mochni, which was similar to a mockingbird. The more Mochni talked, the more he convinced everyone of the differences between them: the differences between people and animals and the differences between people themselves, due to the colors of their skins, their speech, and their belief in the plan of the Creator.[7] The animals drew away from people.

Also among them was the handsome Ka'to'ya, who took the form of a snake with a big head.[8] He led the people still farther away from one another and their pristine wisdom. The people became fierce and warlike; there was no peace. The Creator and So'tuknang decided to destroy the world and start over.

Those who remembered the plan of the Creator were led to a big mound where the Ant people lived, and there they were kept safe when So'tuknang destroyed the First World by fire.

Today, a number of excavated underground ancient cities have been discovered in Turkey. One, called Derinkuyu, is several stories deep and could have held as many as twenty thousand people, along with livestock and food storage. Perhaps the "Ant people" were people who lived underground like ants. Sometimes ideas in myth seem so strange we think they must be fiction. Then archaeologists discover something like these ancient underground cities, which had wineries, stables, and up to five hundred–kilogram round doors to close off the outside world — an ideal home for "ant" people.

When the First World cooled off, So'tuknang created the Second World. He changed its form completely, putting land where the water had been, and water where the land had been. This description concurs with what we know of geology. The movement of the continents caused dramatic shifts in landscape, so that in the southwestern United States today, what was once tropical forest with dinosaurs grazing along the ocean is now the Rocky Mountains and the Colorado Plateau, the regions where the Hopi reside.

So'tuknang thanked the Ant people for saving his people and told them to go into the Second World and take their place as ants. In the Second World, the people were close in spirit and communicated from the center at the top of the head, and they sang joyful praises to the Creator. They did not have the privilege of living with the animals; the animals were wild and kept apart. Being separated from the animals, the people tended to their own affairs. When they began to trade and barter, trouble started.

The more the people had, the more they wanted; they began to fight, and wars occurred. Again, the world was destroyed; and again, the Ant people protected the chosen people by allowing them into their underground world. This time the twins were told to leave the poles, and the earth spun crazily and froze into solid ice.

Many years passed and the people lived happily underground. Then the Third World was created. The people advanced rapidly, creating cities, but again, they eventually became wicked, killing each other. This time, the world was destroyed by water. Continents broke apart and sank to the bottom of the seas.

In the Hopi story, the current world is the Fourth World.

By one count, 272 cultures describe a great flood destroying the world, including the Hopi story of the end of the Third World.[9] In many of these stories, the creator instructs people to save animals, as in the story of Noah's Ark, and in others, animals are involved in saving humans, just as the Ant people saved the Hopi. Other examples include the birds in the Bible story searching for land and a great fish warning the first Hindu man, Manu, of the impending flood. The pervasive nature of the great flood myth, and of our cooperation with the animals to survive, supports the notion that this story represents a global Truth.

Numerous cultures also describe other times of destruction, such as fire and polar shifts, in which most of life perished. Modern science concurs with fossil evidence that indicates there have been at least five distinct mass-extinction events. During these periods, a significant fraction of all the species on earth became extinct in what was essentially a geological instant.[10]

These mass extinctions occurred before the time we currently believe the first humans inhabited the earth, so it is even more striking that human stories seem to contain a "memory" of them. Might there also be truth to memories of a time when humans and animals communicated easily and lived in peace? Some spiritual teachings believe humans existed on earth long before our current geological records show. Who knows what lies buried below, in depths where we have yet to dig? Archaeologists find new fossilized remains on a regular basis, opening up the possibility that some fantasy-like stories may contain more truths than we currently recognize.

Australian Aborigines

The following creation and paradise story of the Australian Aborigines comes from Robert Lawlor's book *Voices of the First Day*.[11] The Australian Aborigines have the longest continuous cultural history of any group of people on earth, beginning about fifty to sixty-five thousand years ago. As with all ancient tales, Lawlor's version is just one version of their creation story.

Australian Aboriginal writer Goobalathaldin (Dick Roughsey) explains that creation on earth began when the first human beings arrived from the stars. They possessed supernatural powers and created the land and sea. Everything was good until floods, volcanoes, droughts, and earthquakes rocked the land. Out of fear, the first ancestors sought refuge in a most remarkable way. They transformed into animals, plants, insects, and rocks. As this Dreamtime creation commenced, the earth became populated with a multitude of life-forms.

The Creative Ancestors were vast, unbounded, vibratory fields of energy. They created with their breath by naming. Just as one creates sounds or songs with the vibration of breath, the Aborigines describe the Dreamtime creation as the world being "sung" into existence.[12]

As the Ancestors traveled across the barren countryside, their travels shaped the landscape. When they slept, they dreamed of adventures for the next day. They dreamed of things and created them: ants, wallabies, emus, crows, lizards, snakes, grasshoppers, plants, and humans. The Ancestors created all these things simultaneously, and each could transform from one to the other. Lawlor writes, "A plant could become an animal, an animal a landform, a landform a man or woman. An Ancestor could be both human and animal." The Ancestors eventually retired into the earth, the sky, the clouds, and the creatures, to reverberate within all they had created. "All creatures   from stars to humans to insects — shared the consciousness of the primary creative force, and each, in its own way, mirrors a form of that consciousness."[13]

The ancestral energy that shaped the earth was referred to symbolically as the "Rainbow Serpent." It resonated in the shapes and lives of the earth as a usable force and nourishing spirit. The Rainbow Serpent represents the electromagnetic spectrum of light, a profound metaphor for the unity between the tangible and the invisible worlds.[14] It connects the earth and celestial realms.

Over vast periods of time, the Rainbow Serpent, like the earth's magnetic field, alternately extinguished and re-created life over the whole earth.

The Aborigines believe the Ancestors created the world perfectly, and it stayed that way so long as humans adhered to the universal law. Abandoning the Dreaming Law forced humankind to leave the garden. Everything changed. The myth of the Southern Cross tells of the first death and how people faced a moral dilemma — whether to kill to survive or die.

The myth relates how drought caused a lack of vegetation, so a man killed a kangaroo rat, which he and a woman ate. A third man refused to eat it and died. A black figure with huge fiery eyes lifted the man into a hollow tree and raised the tree into the southern sky; following it were two yellow-crested cockatoos.

The tree planted itself near the Warrambool, or Milky Way, which leads to where the sky gods live. The tree gradually disappeared from their sight until all that remained were four eyes: two were the eyes of Yowi, the spirit of death; the others were the eyes of the first man to die. The pointers were the cockatoos. These stars make up the constellation of the Southern Cross, a reminder of the first death.[15]

"The Australian Aborigines speak of *jiva* or *guruwari*, a seed power deposited in the earth," writes Lawlor. Every meaningful life process or event that occurs leaves behind a vibrational residue in the earth, just as plants leave an image of themselves as seeds. A seed vanishes the moment it germinates, becoming a plant. At this moment, the seed's latent power springs into action. The seed dies in order to physically manifest, whereas the plant manifests and then dies, leaving seeds. The "seed of the archaic" is maintained in the universal myth of the Golden Age.[16]

The spiritual nature of animals remains today as it was in the beginning, a representation of the Creative Ancestors. The vibratory serpent energy that created the many forms lives on as platypus, plant, and person. The seed power and Rainbow Serpent correlate to a description of DNA. Deoxyribonucleic acid emits photons at the visible spectrum of light. This double-spiral, serpent-like, molecule carries the genetic information or seed power present in every cell of every living thing, including bacteria, broccoli, and bison.

Perhaps these ancient tales describe something we understand today with modern science. Anthropologist Jeremy Narby calls DNA the ancient energy of the creator.[17] Indeed, the art and stories from many ancient cultures relate the same idea. Images of twin creator snakes appear in art from a Mesopotamian seal dated circa 2200 BCE. Aztec art from 400–600 BCE depicts DNA-like, double-helix serpents, and the Aztec god Quetzalcoatl translates as "serpent" and "twin." Ancient Greek stories describe Zeus as a serpent; Scandinavian rock art depicts serpents with a cross; and in ancient Egypt, a serpent crown found on a mummified pharaoh depicts two conjoined serpents, and the serpent symbolized the beginning and end of time. Paintings from Peruvian shamans depict images that resemble snakes containing chromosomes. All appear synonymous to rock paintings of the Australian Aboriginal Rainbow Serpent from six to eight thousand years ago. Perhaps Narby, the Peruvian shamans, and Australian stories are true: The creator "God" vibrates in every living thing, actively creating through the physical wisdom embodied in DNA.

Ancient Hebrews

Chapter 7 discusses the Hebrew and Christian beliefs about the spiritual nature of animals in more detail. Here, I briefly discuss the biblical creation stories.

As in the Hopi and Aboriginal stories, creation begins with sound as the "word" of God. According to the Gospel of John, "In the beginning was the Word, and the Word was with God, and the Word was God."

Two different versions of the creation story appear in the Book of Genesis, chapters 1 to 3. In Genesis 1:20–21, the animals are made first: "And God said, 'Let the waters bring forth swarms of living creatures, and let birds fly above the earth across the firmament of the heavens.' So, God created the great sea monsters and every living creature that moves, and with which the waters swarm." Then God created a human male and female in his own image, and gave them dominion, or control, over the fish, the birds, and every living thing on earth.

In Genesis 2, man is made first from the dust of the earth and then the animals are made in the same way. The word comes in the form of naming. As Genesis 2:19 says: "So, out of the ground the Lord God formed every beast of the field and every bird of the air, and brought them to man to see what he would call them."

Regarding food, in Genesis 1:30, God said, "And to every beast of the earth, and every bird of the air, and to everything that creeps on the earth, everything that has breath of life, I have given every green plant for food."

Please note the description of living creatures having the "breath of life." Biblical scholars consider this breath to refer to the eternal spirit of a being.

Some people believe that animals have lower souls but not eternal spirits, because only Adam received the "breath of life" from God, according to the Bible. However, animals breathe and are alive, and as this verse indicates, every beast, bird, and creeping creature has the breath of life. The website Bible Hub provides twenty-one biblical translations of Genesis 1:30, and about half clearly support the notion that animals have the breath of life; three of them replace the phrase "breath of life" with "living soul." Others use wording such as "wherein there is life." Debate arises as to whether animals received the "breath of life," the omnipresent, eternal, transcendent spirit of God, and I discuss this debate in chapter 7.

To continue the creation story, although the human male and female were both naked, like the animals, they were not ashamed. However, God warned, "You may freely eat of every tree of the garden; but of the tree of the knowledge of good and evil you shall not eat, for in the day that you eat of it you shall die" (Genesis 2:16–17).

Later, the serpent appeared — a being more "subtle" than the other wild creatures — and he said to the woman, "You will not die, for God knows that when you eat of it your eyes will be opened, and you will be like God, knowing good and evil" (Genesis 3:4–5). So Eve ate the fruit and shared it with Adam. Then they realized that they were naked; they became ashamed and covered themselves.

This is a powerful description of the birth of duality, of the concept of right versus wrong, of good and evil, and of humanity's first moral judgment and the pain of shame.

As Genesis 3:22 says: "Then God said, 'Behold, the man has become like one of us, knowing good and evil.'" Genesis 3:23: "therefore the Lord God sent him forth from the Garden of Eden, to till the ground from which he was taken."

Before they ate from the tree of the knowledge of good and evil, humans were innocent and one with nature. Humans left paradise when they gained moral judgment. Before that, Adam and Eve only knew life — they experienced monism, or a sense of oneness with existence. After eating the fruit of the tree of good and evil, they gained dualistic thinking — they distinguished right from wrong, and humans separated from nature, just as when the Talker in the Hopi story convinced people of the differences between themselves and the animals. The birth of duality created ego-centered shame, which leads to death. According to psychiatrist and teacher of enlightenment David Hawkins, shame "is perilously proximate to death."[18] With shame comes blame, guilt, and vindictive hate. It causes such agony that violence or neglect follow, either to self or others, including suicide and murder. Once humans gained dualistic thinking from the tree of the knowledge of good and evil, humanity began to experience death.

The forbidden fruit is another common theme across cultural myths. Humans and animals know only life in the garden until the first death or the consumption of some forbidden food associated with the threat of death. The first Burmese man consumed a particular kind of rice and became so gross and heavy that he was unable to ascend to heaven. Fruit was the offending substance for the Kalmucks of central Asia, for Jews, Christians, and Muslims, as well as for the Masai of Tanzania. In Greek mythology, the forbidden fruit is comparable to Pandora's box, which contained all the evils of the world.[19]

Richard Heinberg writes, "In nearly all languages, the word *fruit* is used metaphorically to refer to the result of a creative process."[20] People are meant to tend to creation, but they give up the stewardship of the garden with an obsessive desire for their own creations in manifested form — forbidden fruit. The physical world also traps people. In the case of Indo-Iranian mysticism, it says that the pure, untainted self — Adam — fell from perfection because of his attraction to earth, the physical.[21]

The first experience of violence also results in our eviction from paradise, as the following African myths show.

African Myths: Barotse, Bantu, and Yao

Here, I review two African paradise stories taken from *Memories and Visions of Paradise* by Richard Heinberg.[22] The first myth comes from the Barotse floodplain of the upper Zambezi River in Zambia.

As the story goes, the creator, Nyambi, lived on earth with his wife, Nasilele. Nyambi created fishes, birds, and animals. But one of Nyambi's creatures was different from all the rest, Kamonu, the first man. Heinberg writes, "Kamonu was special because he imitated everything Nyambi did. If Nyambi made something out of wood, Kamonu would do the same; if Nyambi created in iron, Kamonu would work in iron, too."

Kamonu served as Nyambi's apprentice, until one day Kamonu forged a spear and killed an antelope. Despite Nyambi's protest, Kamonu continued killing. Nyambi realized that he had lost control of his creature and grew angry. "Man, you are acting badly," said Nyambi to Kamonu. "These are your brothers. Do not kill them."

And so, Heinberg says, "Nyambi drove Kamonu out of Litoma, his sacred realm, but Kamonu pleaded to be allowed to return. Nyambi gave the man a garden to tend, hoping thereby to keep him happy and out of mischief. But when a buffalo wandered into Kamonu's garden at night, he speared it, and when other animals came close, he killed them, too.

"But after a while, Kamonu discovered that the things he loved were all leaving him: his child, his dog, and his pot (his only possession) all disappeared. He went to Nyambi's sacred realm to report what had happened, and there he found his child, dog, and pot. They had fled Kamonu and returned to their real home."

Kamonu asked Nyambi for the power to keep his things — with no intention of changing his murderous behavior, the real cause of his losses, but Nyambi refused.

Meanwhile, Kamonu's descendants spread over the earth, killing animals. So Nyambi decided to move away from the earth altogether, and he ordered a spider to weave a web to reach an abode in the sky for Nyambi and his court.

The second African myth is from the Bantu and Yao of equatorial southern Africa.

In the story, the animals watch the people rub two sticks together and make fire. Here's one description of what happens next:

> The fire caught in the bush, roaring through the forest, and the animals had to run to escape the flames.
>
> The people caught a buffalo, killed it, roasted it in the fire, and ate it. Then the next day, they did the same thing. Every day they set fires and killed some animal and ate it.
>
> "They are burning up everything!" said Mulungu, the creator. "They are killing my people!"
>
> All the beasts ran into the forest as far away from humankind as they could get. "I'm leaving!" said Mulungu.[23]

Humans tend to be more creative than other animals. Beavers make dams and birds make nests, but humans make hydroelectric plants and airplanes. These African myths highlight the creative nature of the human being. Humans take after the creator; we love to make things, and we become very attached to our creations. The ego is our personal identity. This idea that we are separate individuals lies in contrast to the monism of the garden paradise. Creativity relates to the ego; it drives us to create things and identify with them.

The first death and killing play prominently in many creation myths; inevitably, the first violent act means the end of paradise. Richard Heinberg explains that in nearly all African myths, God leaves humankind because of our cruelty, quarrelsomeness, and insensitivity to nature.

Theologian L. Robert Keck describes the beginning of violence as an evolution of the human soul. He sees the garden paradise as "Epoch I" of soul development. Humanity's childhood is characterized by unity with nature, respect for animal powers, a nonviolent relationship among people and animals, and a focus on the feminine.

As the soul of humanity developed, humans acted like adolescents. They separated from the creator and gained immature notions of power and control. The social structure became patriarchal, and violence began. Dualism and reductionism were born during Epoch II; humans separated wholes into parts and binged on analysis.[24]

According to Keck, Epoch III is the less-violent future we are gradually entering.

Animal Symbols

Several animals play prominent roles in creation stories. Consider the symbolic meaning of the snake. According to the Native American Black Elk, "Any man, who is attached to the senses and to the things of this world, is one who lives in ignorance and is being consumed by the snakes, which represent his own passions."[25] According to the Hopi, the snake is the symbol of Mother Earth; similarly, the Sumerians believed the serpent represents the power of the Great Mother.

Other native teachings give the snake the power of creation, sexuality, and transmutation. The skin-shedding snake is considered a symbol of rebirth and transformation. For the Hindu, Kundalini energy is envisioned as a serpent that symbolizes sexual, creative energy. The Rainbow Serpent is the Australian Aboriginal symbol of the earth-uterus, or "universal energy." Perhaps the twins mentioned in the Hopi myth represent the double-helix molecule DNA, which exists in all creatures from ants to zebras. In Greece, among other places, the snake symbolizes healing. The depiction of the snake in medical emblems comes from the biblical story in Numbers 21:8–9: God tells Moses to put a serpent on a pole so that anyone who is bitten by a snake can look at it and live. In Egypt and China, the snake is the symbol of inner knowing and clairvoyance — a subtler being, indeed.

Perhaps the wounded snake I saw in Dawn's driveway symbolized my own passions, my obsession with healing and an immature notion of control, as well as my own needed transformation.

The spider also appears commonly in creation stories as another symbol of creativity and as the weaver of fate. To Native Americans, she is the grandmother, the benevolent Earth Goddess, the link to the past and future. She spins her web to catch her prey, much as humans are caught in the web of illusion (that is, the physical realm). The spider remains entangled in its web for food and survival as humans remain involved in their earthly affairs. To become enlightened, humans must detach themselves, as Hindus and Buddhists believe. The process of death is sometimes described as a veil or web being removed.

According to the myths, when paradise was lost to humanity, we gained dualistic thinking, moral judgment, and shame, but what about the animals? Are they still innocent, or do they have moral judgment? Although animals fight over territory, I wonder if nonhuman animals possess the same dualistic ego as human beings. Science has no way of telling us; we can only observe them and ponder their behavior.

Iris in Paradise

Mid-June, in La Plata County, is all about foals. The snow melts off the fields, and baby horses sleep in the sun next to their grazing dams. Horse doctors drive from one ranch to another attending to sick neonates and helping mares get pregnant again.

It was a day in mid-June when I met a foal that I later named Iris. I was in the truck, as usual, when a client of fifteen years, Carmen Sherman, called.

"Silver just had her foal, and, well, I don't know what to think. It looks like she doesn't have any eyes. It's just red where the eyes should be. Something's wrong with her eyes. Can you come see her?"

I was en route to another ranch, but I detoured immediately and drove directly there. As I approached the chocolate-colored filly, I could see that Carmen was right. "Let's tie Silver up so I can get a closer look."

Carmen haltered Silver and tied her to the rail. The mare roared in frustration, calling to her foal, throwing her rear from side to side, stomping and stirring up the dust. Her tail wrung as she grunted and squealed, kicking out in frustration like a good mother being violently protective.

I grabbed the foal around the chest and rump, being careful not to get between her and Silver. Carmen stepped over to hold the neonate, one arm around the chest and one around the rump, while I looked at the orbits. They were much smaller than normal, with no ocular tissue at all in the right socket, just pink conjunctiva. The left opening contained a small amount of something that looked like a lens, smaller than a marble.

"Anophthalmia, no eyes," I said. "I've never seen this before." My mood sank to a quiet sorrow. I knew I would have to kill her. My emotions stirred memories of other deformed babies. Carmen started asking questions, but I was somewhere else, remembering a similar situation with another genetically imperfect neonate I once had to kill.

A baby llama was born with choanal atresia, no opening between the nose and throat. She was beautiful and lively but unable to breathe when she nursed — a condition that would eventually cause her to die slowly by starvation. Another unfortunate factor on commercial breeding farms is that they cannot afford to keep animals with genetic defects around that might deter potential buyers. In animal husbandry, one culls the unhealthy as nature would.

The llama owner could not bear to participate in the killing. So I had to steal the baby, called a cria, from her mother, restrain her myself, and inject a fatal solution into her vein as she struggled against me. I hated myself.

Now, the thought of killing this neonatal foal, as her mother screamed, was too much for me to concentrate on Carmen's questions.

"You can untie Silver; let's get out of here and leave them alone. Has she nursed yet?"

"No, she's having trouble finding the nipples."

This was Silver's first foal, and she obviously liked her baby. She constantly nickered to her, sniffed and nuzzled her, playing interference between the filly and any human who dared to get near. Silver's mother, Goldie, and her sister, Copper, were watching from the neighboring stalls with their healthy foals at their sides.

"Goldie has been really upset about this filly," said Carmen. "She acts like she knows something's wrong."

As Carmen tried to explain about her old mare's behavior, her words set my mind wandering again, this time about Goldie. She had delivered six foals, including Silver and Copper. She had lost one at birth. Carmen and her husband, Ben, had driven into the barnyard on a spring afternoon to find the entire herd of horses racing around, calling out in distress, upset about something. They walked into the paddock to see what all the fuss was about and found a placenta. Goldie must have had her foal earlier than expected, but there was no foal. Ben searched outside the paddock and found the neonate along the irrigation ditch. It had fallen in the ditch, been swept downstream, and drowned.

Afterward, Goldie would not get pregnant the rest of that year. Every year before, she had given birth and conceived again on her first breeding. But not that year; we tried everything medically possible. The vet at the stallion's barn and I both appeared incompetent. My diagnosis — her heart was not in it. We waited a year, and she conceived without difficulty, and she had every year since.

Now, I wondered how this foal's death would affect Silver.

Carmen asked, "What do we do?"

We both knew that foals get most of their protective antibodies from the colostrum, the first milk. Without suckling, this foal would soon get weak, hypoglycemic, and hypothermic. "She'll probably die tonight. It's been cold, and she is having trouble finding the teat. Let's let nature take its course."

Carmen agreed. As I explained to her, no one knows what causes anophthalmia. It may be genetic, inherited, or it may be congenital, developmental, from a virus or toxic plant. It's extremely rare. In case it is genetic, it's best not to rebreed the same mare to the same stallion.

The next day I called, hoping to hear the filly had died, but Carmen said, "She's doing great. She's nursing and running around. Now what do we do?"

We discussed the options of letting her live, who might want her, and such. That gave me another day to avoid the inevitable. In a way, it felt cruel to leave the blind baby stumbling about a rough environment, injuring herself. There was also the fact that Carmen and Ben bred horses for sale, and nobody wanted to see a foal with a genetic defect. We eventually came to the same conclusion.

The day came. The thought of killing Silver's foal made me sick to my stomach. My only consolation was the thought that Silver was a happy mother for a few days. On my way over, I stated out loud, "I want this to happen as gracefully as possible."

Ben was waiting for me, seated at the barn door. The big, black mustache covering his lips did not conceal his frown. "This sucks," said Ben. "She's healthy and beautiful in every other way."

I looked to see an active filly bouncing around her mother, nudging and nursing from her. The mare seemed peaceful and happy, watching her baby and nickering softly.

"What's that?" I asked about a big red welt on the filly's chest.

"Oh, that's one of the places where she's banged herself bumping into things."

Carmen joined us. "Are we doing the right thing?"

I shook my head and sighed. "I don't know."

We silently went about the task. Carmen held Silver by the lead and offered her a large tub of grain. Ben grabbed the filly — which I named "Iris" because of her lack of ocular irises, and because it is a pretty flower that doesn't bloom for very long. The mare gave a concerned nicker, looking to follow the foal. Carmen shoved the grain in her face, and she ate eagerly. Ben carried the filly into the adjoining stall, where I waited, and closed the gate.

Iris squirmed once against the injection and then died without the mare ever noticing. We laid her on the ground and walked out of the paddock, leaving the gate open. The mare finished her oats and came searching for her baby. She sniffed at her and seemed content to find her sleeping, then went over to eat hay.

I took a deep breath, gazing off toward the river, and caught Goldie staring at me. She was in a pen down the hill, watching me with her neck stretched up high and resting her jaw on top of a five-foot-tall panel fence. All the other horses were busy nibbling. But Goldie was staring at me. "Hi, Goldie," I said. Her eyelids slowly closed then reopened with her eyes focused on me.

Goldie knew exactly who I was and what I did, and I knew her. I had floated her teeth, flushed the lacrimal ducts of her eyes, and been inside her orifices at the other end many times. I had cared for this mare both when she was well and when injured and in pain. She knew my touch and I knew hers. As we stared into each other's eyes, we felt connected. We sensed each other's thoughts. She was not upset, anxious, or angry. Her glare was the calm, stern, but approving look of a matriarch supervising chores. In the wild, I thought, the duty of culling the unfit might have been hers before wild animals intervened and ate the filly alive. It would have been safest for the herd to leave that foal behind.

Carmen had mentioned how Goldie acted as if she knew something was wrong with Iris. Maybe Goldie had an opinion or judgment about the foal. Animal behaviorists might call her behavior "instinctual," which can be defined as an inner guidance or intuitive power, though not necessarily associated with mental evaluation. Instinctual acts are more like my self-defense moves in karate, without thought. Still, I wondered what motivated her behavior.

Carmen told me later that Silver nudged and pawed at her foal many times during the night. By morning, when they opened the gate, she ran out to the pasture to eat grass. She whinnied to the stallion and trotted around with her tail up, then went on grazing.

Other horses have grieved longer, as much as a week or more. Maybe grief had an effect on Goldie and her inability to conceive after her foal drowned. But Silver moved on with her life. She did not act depressed or exhibit shame about her inability to create a healthy baby; she appeared to hold no guilt for not protecting it. She showed no hatred toward people.

Do horses and other animals make moral judgments that include blame, shame, and guilt? We know they have feelings and care about one another; they have emotions and opinions. Mostly they are focused in the present, but they also remember and anticipate. They have instincts and guidance that directs them for migration, or awareness of impending weather. They make decisions based on preference, but perhaps they do not dwell in moral agony over what is good or evil. However, after considering Goldie's emotional response to losing her foal and to the birth of Iris, I wonder. Events may not feel right, and animals may sense when something is wrong.

Historically, some animal behavioral scientists state that animals are amoral because they are not self-aware. Self-awareness means knowing that you are separate or different from another — that every individual has unique mental states and emotional experiences — and that you can assess the future or reflect on the past. I suspect animals have self-awareness, yet perhaps they have some other awareness, a sensation of oneness. If animals are not aware that they are separate, then maybe they are not. Perhaps they still live in monism in the garden and know we are all one. Only humans assume separation.

Some people say that having morals makes us superior to animals. But we are too often stuck in grief, reflecting on the past, trying to find something to blame. We feel bad while the horse is enjoying life. The animal's way seems more paradisal. But what do I know? I may be as blind as Iris. As David Pugh, another large-animal veterinarian from the University of Georgia, often says — "I don't know. I drive a truck for a living."

CHAPTER 3

The Hunter-Gatherer: Shamanism

In the beginning of all things, wisdom and knowledge were with the animals; for Tirawa, the One Above, did not speak directly to man. He sent certain animals to tell men that he showed himself through the beasts, and that from them, and from the stars and the sun and the moon, man should learn. Tirawa spoke to man through his works.
— CHIEF LETAKOTS-LESA[1]

Outside the garden, separated from our spirit home, humans had to work for food. Hunter-gatherer societies emerged, yet remaining within these groups were individuals, shamans, who retained paradisal skills, such as crossing over to the spirit world, flying, communicating with animals, and transforming into animals. The average person developed conflicting ideas — about good and evil, right and wrong. Shamans lived outside this duality and retained the ability to "transcend opposites, to abolish the polarity typical of the human condition, in order to attain to ultimate reality," according to expert Mircea Eliade.[2]

To manage the presence of death, the shaman was a healer, a magician, and a psychopomp, one who conducts spirits or souls to the other worlds. Thus, the shaman was, and is today, an integral part of tribal communities.

In general, a shaman defends life, health, fertility, and the world of "light" against death, disease, sterility, disaster, and the world of "darkness."[3] He or she is a prominent personage who, symbolically or in reality, moves between the physical and spiritual realms.[4]

Shamanic affairs are similar throughout the world's various shamanic cultures. Tribal beliefs describe a spirit reality from which we came and to which we can return. The shaman first learns to reach the spirit realm through an initiation, which involves a ritualistic death, dismemberment, and rebirth in shamanic form. During "death," the person experiences an ecstatic journey with transformation and transportation to the other world, thus sacrificing the profane physical condition.

The shamanic practice is a "technique in ecstasy," the ultimate religious experience, according to Mircea Eliade. The shaman masters a state of ecstatic joy, and in this mystical state, he or she attempts to transcend the human condition, to reach the source of spiritual existence. Eliade calls the experience of ecstatic joy "nostalgia for paradise," a pervasive spiritual/religious technique found across the globe and characteristic of the mystical experiences of many religions, including Christian, Hindu, Buddhist, Sufi, and tribal.[5]

The ancient shaman's activities often involved animals, largely because of the need to hunt. The shaman conversed with the guardians of the game, spirits in animal form, and requested that creatures be provided for food. Primitive hunters believed that nonhuman beings were similar to humans, except that the beasts had supernatural powers. They could change into people and vice versa. Mysterious relationships existed between individuals and certain animals who were guardian spirits. For example, a guardian bear spirit might give protection or guide the person in the use of herbs.

Animals, Spirits, and Shamans

Three characteristics of shamanism pertain to animals — the shaman can fly, communicate with animals, and transform into animals.

FLIGHT

Mircea Eliade explains how shamans retained the ability to fly. "According to many traditions, the power of flight extended to all men in the mythical age. All could reach heaven, whether on the wings of a fabulous bird or on the clouds."[6] The shaman retained the primordial nature of flight and transcendence of the human condition to visit the spirit realm. Others only reached heaven at death. The Egyptian Book of the Dead describes the deceased as a falcon flying away. Flight is an integral part of many theological ideologies, with similar symbolism across religious teachings. For example, the Hindu text *Pañcaviśma Brāhmaṇa* says, "Those who know have wings."[7] As the Judeo-Christian angels have wings, so does one of the most important symbols of the Native American Sioux religion, the great Thunderbird. In *The Sacred Pipe*, Joseph Epes Brown wrote: "He is the same as the great one-eyed Bird, *Garuda*, of the Hindu tradition, or the Chinese Dragon (the Logos), who rides on the clouds of the storm, and whose voice is the thunder. As giver of Revelation, he is identical in function to the Archangel Gabriel of Judaism or Christianity — the *Jibrail* of Islam."[8]

Other examples include Odin of Germanic mythology, sometimes called "Eagle." Greek sorcerers also professed to furnish the souls of the dead with wings to fly to heaven, and Buddhists discuss two forms of flight, mystical flight and magical flight (which is an illusion).

Mystical flight is also demonstrated by the levitation and flights of saints in both Christian and Islamic traditions. The Roman Catholic Church records as many as seventy levitations by saints, including St. Joseph of Cupertino and Sister Mary of Jesus Crucified, who were both seen flying up into trees. Because shamans enjoy the spirit condition, they can fly to the World Tree and retrieve "soul-birds." A bird perched on a stick is an ancient symbol of shamanism.

Vestiges of shamanic practices remain among the world's religions today. The ancient, essential features of shamanism include the ascent to heaven, the descent to the underworld, and the evocation of the "spirits," all common religious concepts.

COMMUNICATION WITH ANIMALS

In the process of initiation among the Eskimo, a future shaman learns a secret language, often the "animal language," which is needed in order to communicate with the spirits and the animal spirits. Animal language is only a variant of "spirit language."

African shaman Malidoma Patrice Somé's grandfather communicated with chickens. Malidoma's grandfather once explained that a rooster had just told the hen to run into the basket of millet belonging to a one-eyed woman and tip it over. Seconds later, the hen did just that, and the flock rushed over to feast. Malidoma's grandfather said that if one wished to understand the chicken's language, one should stop eating chickens.

A spirit may take the form of an animal and appear to humans in normal waking consciousness. A shaman communicates with several types of spirits in animal form: familiars, helpers, and guardian spirits. There are also more powerful spirits that are tutelary spirits. Native people teach that the universe is a mirror. Any animal or anything else one sees is a reflection of themselves. Every animal one sees brings a message. Communication also comes in the form of visions involving animals.

Altered states of consciousness, such as visions, trances, and séances, are the hallmark of shamanism.[9] In the trance state, the shaman understands all of nature. Trance states are attained by numerous methods: fasting, meditating, drumming, dancing, singing, and many other rhythmic activities, as well as while using psychotropic drugs or during a state of severe illness. Daydreams, night dreams, and lucid dreaming are other ways a shaman communicates with the spirit worlds. An accomplished shaman does not need an altered state and may connect with many realities at will.

Peruvian shamans claim the hallucinations they have after drinking a beverage made from the ayahuasca plant are the way the plant communicates with humans. One shaman said, "That is how nature talks, because in nature, there is God, and God talks to us in our visions."[10]

Nature spoke to Black Elk, an Oglala Sioux shaman, in a vision of "twelve buckskins [horses] all abreast with horns upon their heads and manes that lived and grew like trees and grasses."[11] These images carry important meaning to the shaman as forms of communication from Spirit. In general, the person seeing the vision must interpret it for him- or herself.

However, some prophetic visions have become part of mystical traditions, as is the case with the Merkabah mysticism (or Chariot mysticism) in Judaism (c. 100 BCE to 1000 CE). Merkabah mysticism spawned from the stories of the biblical prophet Ezekiel and his visions. Ezekiel 1:5–10 says: "And from the midst of it came the likeness of four living creatures. And this was their appearance: they had the form of men, but each had four faces, and each of them had four wings. Their legs were straight, and the soles of their feet were like the sole of a calf's foot: and they sparkled like burnished bronze. Under their wings on their four sides they had human hands. And the four had their faces and their wings thus: their wings touched one another; they went every one straight forward, without turning as they went. As for the likeness of their faces, each had the face of a man in front; the four had the face of a lion on the right side, the four had the face of an ox on the left side, and the four had the face of an eagle at the back." For these mystics, the visions in Ezekiel describe the angels, chariot, and throne of God.

TRANSFORMATION INTO ANIMALS

When a shaman transforms into an animal, he or she reestablishes the condition that existed in mythical times before the separation of the human, animal, plant, and mineral worlds. Remember, no veil exists between the physical and spirit realms for the shaman, so the act of moving between the forms of beings is as natural as it was for the Australian Aboriginal ancestors.

The Buryat (Siberian/Mongolian) shaman's tutelary animal enables the shaman to take another form, an animal form. A shaman may put on animal hides or masks to initiate the process. Entering the ceremonial costume facilitates the shaman's contact with the supersensible world. Sacred costumes persist today, even in the robes of priests, which resemble birds. The magical power of flight is assumed by wearing an eagle feather. In general, the shaman becomes what he or she displays, and the shaman may call like a bird or squeal, grunt, whinny, bellow, growl, and make other animal sounds and movements.

In the scientific model of trance, transformation into an animal is documented by researchers. According to Jean Clottes and David Lewis-Williams in *The Shamans of Prehistory*, "A Westerner experiencing an altered state said, 'I thought of a fox, and instantly I was transformed into that animal. I could distinctly feel myself a fox, could see my long ears and bushy tail, and by a sort of introversion felt that my complete anatomy was that of a fox.'"[12]

Shamanic rock art depicting part-human, part-animal beings is found in caves in South Africa, France, Spain, and the United States, of which one can find excellent photographs in *The Shamans of Prehistory* by Jean Clottes and David Lewis-Williams.

A Vision in the Bisti Badlands

I was riding my donkey, a little black-and-white pinto, in the Bisti Badlands of northwestern New Mexico, when a strange vision appeared. The Bisti is a barren, eroded, gray-on-gray desert with a few strands of maroon, rust, and tan. Two lady friends rode their mares along with me. One had ridden in the Bisti Badlands previously and mentioned that the last time she was there, she saw a naked man. We headed east along a thin stream of brackish water where my friend's dogs played in the mud and sand. As the two riders in front of me turned south, I saw before me, in the distance, a man stand up from his seat on a rock and scurry off. He wore a straw hat and a day pack, but no shirt or pants. We were not sure what to make of him. The Bisti is Navajo tribal land, and the terrain is stark and lifeless. It seems otherworldly, perhaps the kind of place a Navajo medicine man or woman might use for ceremony, and we supposed this to be the case with the naked man. We rode by tall steely hoodoos, large round mounds, bluffs with deeply, eroded edges, and large logs of ancient, petrified wood. My donkey seemed disappointed by the lack of green vegetation as she tasted a dry briar and spit it out.

In the hot, midday sun, on this desolate terrain, imagine my awe when above us silently soared a white owl! It flew from behind me about ten feet overhead. The face was not visible, but I could see three sides, and they were mostly white. The fan-shaped tail had faint brown stripes and spots. I know birds. This was no hawk. My two companions also saw the owl, but I was most stunned by its presence. It looked like a snowy owl, which seemed outrageous because they breed in the arctic tundra. We spent some time discussing what it meant to see a white owl in the Bisti Badlands. We understood the belief among the Navajo that an owl is a bad omen, and we agreed not to label the sighting as negative. After all, it was white; it had to be a positive vision.

Shamanism teaches that each animal we see is a reflection, and the observer must interpret each vision they see, but I wrestled with my scientific mind to explain the bird's presence. Spiritual beliefs from other cultures teach that the owl signifies wisdom, or seeing in all directions without illusion. Since the location seemed like the kind of place a shaman might visit, and with no logical explanation forthcoming, my friends and I concluded that the naked man was a shape-shifting shaman who became the owl.

My scientific mind wanted more concrete answers, although I struggled to find them. The only species of owl that can sometimes be albino and might be found around the Bisti is a barn owl. A Bisti wildlife biologist agreed that even the sighting of a white barn owl would be very unusual. Regardless of any "real" explanation, the sighting presents a mirror for self-reflection. To help me find the answer, I consulted numerous spiritual teachers.

Dana Xavier, my clairvoyant friend, said I had acquired a state of consciousness that allows me to see things in the spirit realm. Two female, nonnative shamans (Stacy Couch, an ornithologist, and another who is a veterinarian) both stressed that the interpretation was mine to determine. The bird might be a glimpse into another realm, and because the owl can see in the dark, it might indicate the ability to see beyond what others might miss. It could be my spirit guide, a benevolent spirit being who has agreed to support a person's spiritual evolution. The term *daemon* also applies, as the Latin word for guardian spirit from the ancient Greek, *dímōn*, means "protective spirit."

In *The Art of Shape-Shifting*, Ted Andrews describes the owl as having increased intuition, vision of things not readily seen, and awareness of spirits. White is the color of spirit, and the owl is often referred to as the "night spirit."[13]

All these explanations seem fitting, but words are difficult to find that adequately portray what the owl means for me. To aid in understanding, visualize a beautiful country scene with green grass, flowering bushes, and fruit trees humming with squirrels, birds, and bees. Now imagine an eraser that removes all life from the picture. Erase away the frills and fancy of transient beauty, and one finds the Bisti Badlands — the bare-naked bones of existence. There, behind the veil of our illusions, one finds the eternal, pure Spirit. The Bisti is like a crack in the veil of the physical world we create with our minds. With nothing close to normal reality to focus on, the Bisti opens up the possibility to see beyond the mirage of our mundane existence and into what shamans, Hindus, Buddhists, people who have had near-death experiences, mystics, and clairvoyants call "the reflection world," the spirit reality behind the material world. For me, the Bisti owl was a reminder that Spirit is always with us, everywhere. When we connect with Spirit, we find guidance, wisdom, intuition, and protection. Spirit is omnipresent; we fail to notice it because we are so focused on our problems and projects — the fruits of our creations. I suspect the naked man was there for the purpose of connecting with Spirit. For me, the white owl symbolized and reflected the pervasive presence of pure Spirit.

Shamans around the world experience the ecstasy of union with Spirit. Two examples of tribal shamanic beliefs about animals include the Oglala Sioux and the native peoples of Central and North Asia.

The Oglala Sioux

Crazy Horse dreamed and went into the world
where there is nothing but the spirits of all things. That is the real world
that is behind this one, and everything we see is like a shadow of that world.
— BLACK ELK

The Oglala Sioux of North America perceive the phenomenal world in a more fluid and transparent way, in which no absolute lines are drawn between animals, people, and spirits. The unifying principal of all things is the *Wakan-Tanka*, which is analogous to the one supreme God of other religions. All things are aspects of the one, connected by the wind.[14] As one shaman is quoted as saying: "The Four Winds is an immaterial god, whose substance is never visible.... While he is one god, he is four individuals.... The word *Wakan-Tanka* means all the *wakan* beings because they are all as if one."[15]

Every part of the one is connected in the dimension of the sacred, and each animal characteristic symbolizes the powers it demonstrates. The buffalo is chief of all animals. She represents the feminine, creating earth, which gives rise to all that is. The bear is knowledge, especially of herbs and roots. The moth is the wind contained by the cocoon.

Prior to the arrival of Europeans in North America, the Sioux did not have a sentimental or romantic attitude about animals. Nor did they view them materialistically. They hunted game for survival, and they viewed the hunt as a religious activity that required ritual preparation. The game was a sacred, power-bearing being. They referred to an animal as a spirit. For many Plains tribes, each animal was a crystallized projection of the abstract spirit. One man called Bear with White Paws said, "The bear has a soul like ours, and his soul talks to mine and tells me what to do."[16]

Other important aspects of the Great Spirit are embodied in spiders and birds. The spider (Iktomi) holds a special place for the Oglala. He can fly through the air on a strand of web, he can glide across the water, and he can walk on eight legs.[17] Since birds are two-legged and can fly, they are considered to be supreme — the most important of all the creatures. To interact with an animal is to communicate with that aspect of the unifying spirit. Black Elk related that "one should pay attention to even the smallest crawling creatures, for these too may have a valuable lesson to teach us, and even the smallest ant may wish to communicate to a man."[18]

When an Oglala has a vision or a dream of an animal, that animal becomes significant in a religious sense for the individual; that animal may be a spirit guide. A vision may come during a time of illness, or one may quest for a vision. The ritual vision quest may involve fasting, walking long distances, and participating in the sweat lodge.

Joseph Epes Brown wrote, "It is through the vision quest, participated in with physical sacrifice and the utmost humility, that the individual opens himself in the most direct manner to contact with the spiritual essences underlying the forms of the manifested world. It is in the states achieved at this level that meditation may be surpassed by contemplation. Black Elk has thus said that the greatest power above all in the retreat is contact with silence.'...For is not silence the very voice of the Great Spirit?'"[19]

A vision of an animal spirit guide may come, or people may turn into animals, and animals may turn into people, other animals, or even plants, such as a sacred medicinal plant. The animal spirit guides may provide the human with special powers. For example, the medicine man or woman may learn the uses of medicinal herbs from the bear spirit.

The medicine man or woman who becomes a healer knows that it is not the person who works the cure, nor is it the animal or the medicinal plant. According to Brown, "These are simply the phenomenal channeling intermediaries through which the intangible spirit-power operates. Black Elk made it clear that the power does not come from him, but from the Power above, which is the source of all powers."[20]

Each class of animal has a guardian divinity that is the mother archetype, which may appear as a phenomenal animal. The spiritual or immaterial quality of an animal is the important thing. As there is no clear line between the worlds of animals, spirits, and people, outward forms shift. Thus, the Nagiya, or immaterial self, of a bear may possess a person when that person wants to have the nature of a bear.

For hunter-gatherer societies in general, animals are powerful, immortal spirits. The soul or life of the animal is contained in the bones, especially the skull. It is from these bones that the lord of that wild beast causes new flesh to grow. So the hunter has reverence for the beasts. He kills the animal that is provided for him by the guardian spirit, as agreed between that spirit and the shaman. The hunter kills only what is needed and never wastes food.

The Tungusic and Buryat Tribes

The word *shaman* is Russian, and it derives from the Tungusic language, which is shared by several indigenous tribes in Siberia and Central Asia, and their religious practices. However, the shaman is not a priest. Much of the tribe's religious life goes on without him or her. The shaman functions as a healer by entering into a trance during which the soul leaves the body to travel to the spirit worlds. The shaman allows healing by capturing the patient's missing soul and returning it to occupy the body. The shaman also accompanies the souls of the dead to the otherworld. The mystical journey is perilous, but he or she is aided by spirit guides and sanctioned by the initiation experience.

Animal sacrifice occurs in the history of all the major religions and is discussed in detail in the next chapter. Siberian shamans participate in the sacrificial ritual for the purpose of healing. Remember, the shaman retains a paradisal, nondual mentality, an understanding that there is no death, only transformation. The shaman is the master transformer who can shape-shift, like the Australian Aboriginal creators, and guide others, including the sacrificial animal, to spirit realm for the purpose of healing. The physical world is just an illusion we made up with our concept of dualism. The tribal shaman travels between both worlds, guiding the sacrificed animal on the journey. He or she participates in the sacrificial ritual but is not involved in the actual killing of the beast. Rather, the shaman is only concerned with the mystical itinerary of the sacrificed animal. The shaman conducts the soul of a sacrificed animal to the celestial Supreme God.

Mircea Eliade explains the shaman's powers: "He foresees changes in the atmosphere, enjoys clairvoyance and vision at a distance (hence he can find game); in addition, he has closer relations, of a magico-religious nature, with animals."[21] The Buryat peoples, who are distinct but related to the Tungusic, tell a story that illustrates this.

> In the beginning, there were only the gods (*tengri*) in the west and evil spirits in the east. The gods created man, and he lived happily until the time when the evil spirits spread sickness and death over the earth. The gods decided to give mankind a shaman to combat disease and death, and they sent an eagle. But men did not understand its language; besides, they had no confidence in a mere bird. The eagle returned to the gods and asked them to give him the gift of speech, or else to send a Buryat shaman to men. The gods sent him back with an order to grant the gift of shamanizing to the first person he should meet on earth. Returned to earth, the eagle saw a woman asleep under a tree, and had intercourse with her. Sometime later the woman gave birth to a son, who became the first shaman.[22]

The father of the first shaman is therefore the eagle; the eagle is the shamanic symbol of the Supreme Being, and they have the same name — Ai Toyon — the creator of light. Legends state that each shaman has a bird-of-prey mother. A Yakut legend states that shamans are born in a giant fir tree in the north. In it, a bird-of-prey mother, with the head of an eagle and iron feathers, lays eggs that hatch into shamans.

The first shamans were "white," created by the gods, and they wear white costumes.[23] The color refers to the types of spirits the shaman is assigned to. The white shaman participates in the horse sacrifice ceremony, where a horse is blessed, then sacrificed, and the shaman takes the horse's soul through the heavens directly to the Supreme Creator to whom the shaman prays. If the sacrifice is accepted, the shaman receives predictions concerning the weather and harvest. He or she also learns of what other sacrifices are expected.

The "black" shamans are a more recent creation and wear a blue costume. The term *black* does not indicate evil but signifies the class of shaman. The black shaman may travel to the underworld, where a dog guards the entrance to the realm of the dead. Going to retrieve the souls of sick people, he or she must cross a dangerous bridge that the soul of a sinner cannot cross. The shaman passes the place where sinners are being tortured and succeeds in entering, despite the dog. The shaman may be there to retrieve the soul of a sick child, for which a costly sacrifice may be required. The soul of another consenting victim may be given. In that case, the shaman turns into an eagle, descends on the victim, and tears out the person's soul. On the return from the underworld, the shaman may come back riding a goose.

Often an ecstatic journey to the otherworld may take place in a vision. The Samoyed shaman sees a helping spirit, a reindeer, taking the adventure as he or she sings about its travels. The shaman's spirit learns of the cause of the illness, and if the Supreme Being (Num) sent the disease, the shaman will not treat it; instead, he or she lets the helping spirit ask Num for assistance.

The Yakut and Buryat shamans fly on horseback at great speed. These beliefs, in relation to flight, are figurative expressions for ecstasy, describing the mystical, superhuman journeys into regions inaccessible to humankind.[24]

Interview with a Shaman

Since shamans live today, we have a rich source of information available regarding the ways of paradise, or ecstatic joy, and the spiritual nature of animals. All one need do is ask a shaman. A genuine shaman is elusive, however — they are not always involved in the normal affairs of modern life, nor listed in the yellow pages or on the internet. As Jeremy Narby writes, "In shamanic traditions, it is invariably specified that spiritual knowledge is not marketable."[25] Certainly, the shaman's work deserves remuneration, but by definition, the sacred is not for sale. The use of this knowledge for personal power defines black magic.

I wanted to talk to a modern shaman, one who does not conduct business with websites and retail stores. I met such a man from my husband, Jean-Luc Boucher, who had been on a vision quest with a shaman named Iron Feathers. Iron Feathers agreed to be interviewed for this book, and we spoke over the phone several times.

When we talked, Iron Feathers described himself as a "nonnative, European mongrel." Without a tribal society to teach him, he had to find his own path to shamanism, which he did by studying with shamans in the United States, Mexico, and Peru. He acknowledged his struggle with the responsibility of his chosen path, which he takes seriously. I asked him to describe the process of becoming a shaman, and after considering my request, he cautiously agreed to talk about it.

"I will do the best I can to answer your question," said Iron Feathers. "And know that this is my opinion, and that the spirits are having a mighty laugh at my expense."

Iron Feathers explained that he had undergone several shamanic initiations, dying and being dismembered three times.

"It wasn't that pleasant," he said. "It is hard to describe because you don't know what's going on. You have the training that tells you that this is initiation, but part of you is dying, being torn apart. You know that you can, in fact, die. There are two entities — 'I' who am going through the process and I who am watching. You leave your body; still, a piece remains in the body. It is nondescriptive, not your soul or astral body. When I have heard people ask about it, we are told that it is not important; it happens.

"After dismemberment, you are rebuilt in shamanic form. You are usually given something, like a crystal, and it's inserted into you. Crystals are the only things that don't change between worlds. Everything else changes over there, in nonordinary reality."

I asked, "When do you see your spirit guides?"

"When a spirit helper comes, it shows itself to you four times. In the otherworld, you may see a bunch of different animals, but only the one you see four times agrees to help you. It shows you how to do things over there, where to go."

"Do you think this is hallucination," I asked, "or do you think what you see is real?"

"I have no problem believing it," said Iron Feathers.

I asked Iron Feathers about the experiences of other shamans, such as Malidoma Somé, who describes seeing spirit beings and communicating with animals in normal waking life, without being in a trance state. Malidoma tells the story of his grandfather, who was pronounced dead in a hospital. Malidoma's father put a hyena's tail decorated with cowry shells in the dead man's hand, and his grandfather got up and walked home, four miles, then he was dead again. They woke him up again later to have a feast before his funeral.

Iron Feathers replied, "Well, depending on who you are, that can happen. A friend of mine spent some time with some yogis in India. They were arguing about something, when their guru returned from the dead to tell them that every one of them was wrong. They all saw him. Of course, Christ appeared to his disciples, but I'm not sure that we're talking about the same thing. There is nonordinary reality. At some point, you are in both worlds. For Malidoma's grandfather, the yogi, or Christ, there is no veil anymore. But for us mere mortals, we have to use a trance state."

I asked Iron Feathers if he had ever transformed into an animal and how that happens.

"I watched a Peruvian shaman turn into a very beautiful white egret," he said. "And I have a drawing made by a friend who watched me turn into a bear. So, from this, I know it is a real thing and not just a figment of someone's imagination. Konrad Lorenz, a Nobel Prize winner in the seventies, put forth the idea that in the motor cortex is an image of the body, and the body tries to conform to that image. Some would call this body mapping.[26] What I believe is that the shaman receives the image of the animal in his motor cortex. From the quantum physics point of view, the body, and all matter, re-creates itself every nanosecond. The energy for this change comes from the electromagnetic field of the universe — the Divine, God, or Universal Life Force. Yogis have talked about this kind of transformation for about five thousand years. So the shaman has the image of the bear in the motor cortex, and in nonordinary reality he has the ability to take that image and re-create a new image of a bear. Because he is of strong enough force, he can maintain the form until he has finished doing whatever is required, and then he can revert back to human form. Sometimes one can imagine the same thing happening when he feels like a bear. Inside he is touching the image but not strong enough to bring it into this reality and maintain it."

"What does it feel like to be in shamanic form?" I asked.

"It's not like in the movies where your face rips off. One minute you are human; the next you're an eagle. One minute you're driving a car; the next you're flying in the sky."

"So shamanic form is your transformed animal form."

"Yes, I assume you feel like the animal feels. So as a bear you feel strong, with a strong sense of smell and poor eyesight."

I asked Iron Feathers to describe the purposes for visiting nonordinary reality and to describe what it feels like.

He said, "The idea of shamanism is to bring back information from the other side to help people — find plants for medicine, learn where the deer are. You find out what your specialty is once you get to the other side. My intention is to help heal people spiritually by putting the spirit in balance. Everything is different over there. A cat in this world may be a human or a tree over there. The problem with a spiritual discussion is that we do not have the words to explain it. We lack the language to define the indescribable."

"So if a cat looks like a human in nonordinary reality, how does its spiritual nature differ from that of a human's?"

"Who says it does? I don't know, maybe it doesn't. We used to play a game, 'hide and seek in nonordinary reality.' Since everything changes over there, you may not recognize your friend or your cat over there, so we give the friend something to help us recognize him, or we hold onto the cat during transformation."

In my study of shamanism, I felt confused about the different kinds of animal spirits: divine animals, deities in animal form, and spirits of animals. I asked Iron Feathers if he could explain the difference.

"A divine animal is the divine in animal form, not the divine in an animal," he said. "Spirit created humans to recognize itself — so they say. The divine in animal form is a projection so that we can recognize it because we can only see what we can recognize."

Once he said this, I understood my Bisti owl a bit more.

Iron Feathers continued, "Humans in turn created spirit so they could recognize it, like trying to find words to explain spirit. For example, Ben Franklin created his own vocabulary for electricity. Electricity always existed, but we did not discuss it until he made up the words to describe it."

"Do you find that there is a hierarchy of animal spirits?" I asked.

"There are animal spirits in all three worlds, the upper, middle, and underworlds. They are mainly in the underworld, playing. One teacher said, 'Go play with them; they want to have fun and be seen.' Animals differ from one plane to the other. And again, we can only conceive of things being the way they are for us. So it makes sense to us that there is a hierarchy like 'mid-level management' because our world is set up that way. What I have experienced is that spirit is spirit, and we're all part of spirit — so there is no hierarchy. There may be different positions of enlightenment, like for the Hindu. You do not go from board member to CEO, but perhaps become more filled with spirit or more a part of spirit. When I am in nonordinary reality, some things are similar to this world, and there are places that have no relation to here and I don't know how to describe them."

Vision Quest

In my efforts to better understand the spiritual nature of animals, I wanted more than to just talk about shamanism and the spirit realm. I was eager to hear the silent voice of the Great Spirit and see what animals would teach me, and so in September 2001, I prepared to set out upon my own vision quest. The plan was to live in the high mountain desert of southeastern Utah alone for three days and nights with no food and only a tarp for shelter. In preparation, I participated in sweat lodges, meditated, prayed, took daylong solo hikes, and wrote in a journal about the animals I encountered.

Several days before my departure, as I drove to my first appointment, I heard shocking news on the radio. I called home immediately to tell Jean-Luc to turn on the television — it was the morning of September 11, and a terrorist attack had turned New York's World Trade Center towers into a pile of rubble. Airports nationwide were closed; terrifying images appeared on the television and the internet; and the radio roared with angry voices. People I spoke with in Durango seemed terrified and left their jobs to be with family.

I prefer the quiet of nature when turmoil strikes, and so I looked forward to my journey a few days later. On September 16, I traveled far from the noise and fear of the city into the wilderness until eventually I sat surrounded by flowers and the soft sounds of insects, birds, and frogs. Warm air, cool breezes, fluffy clouds, autumn colors, and solitude welcomed me. Far from the news broadcasts, I found peace, while the rest of the nation suffered from sadness, worry, and fury.

My guide was CJ, a petite, older woman, and we made the trip together. We hiked down into a canyon to camp and prepare for my solo trip into the wilds. Bear scat and mangled chokecherry bushes along the trail reminded me of the serious bear problems in Durango, where dumpsters and orchards entice them. Drought conditions had created a wild food shortage. Bears raided people's trash bins, ravaged fruit trees in yards, and even went into some houses. But we did not talk about that. I made every effort to follow CJ's instruction to leave the world behind and be fully present in the moment.

As we descended into the ravine, the cathedral-like peaks on the opposite wall came into view. The steep, rocky cliffs with scattered brush looked like the perfect place for a mountain lion to rest and groom as she overlooked her territory. On the valley floor below, we set up base camp under two tall ponderosa pines.

The next morning CJ counseled me. She lit a bundle of white sage and smudged me with its smoke. Then she sent me out alone to find a place to live for my three-day vision quest.

I hiked along a faint trail following the stream as it curved through the canyon. Thick vegetation underfoot indicated that no person had walked on it for some time. The perfect location took some consideration, but I finally found my spot about a mile away from base camp. The home I chose was near the water, with shade. I hiked back to camp to get CJ and take her to see the place, so she would know where I was. Then we went back to camp for our final meal together.

"Get your pad and journal. Before you go out tomorrow morning, it may be helpful for you to choose a power animal to represent each of your chakras," said CJ. Chakras are vibrational energy centers along the central axis of the body. The word *chakra* is Hindu, but the centers are common in many religious beliefs. They are perhaps related to the endocrine glands. CJ instructed me, "Lie back and, as I ask the question, choose an animal, the first that comes to your mind."

I reclined on my pad underneath the two large ponderosas, enjoying the soft pine-needle-covered ground and the scent of vanilla from the tree bark.

"Relax and look into your first chakra, the area at the base of your spine. This is your connection to the earth. What animal do you see? If you wish, give it a name."

The first chakra animal was "Sheba," a cat I once had who used to hike with me. With each question CJ asked, I imagined an animal and drew its picture. For the second chakra, the personal power center, I chose a reptile and drew the image of an alligator curled into a circle in my abdomen, representing qi circles, a way of moving energy in the pelvis. A bumble bee playing in the flowers was the third chakra animal. For the heart center chakra, I chose "Horse runs free — Liberty." For the fifth chakra, the throat, the area of speaking the truth, I chose the owl.

"Take a breath and relax as you look into your sixth chakra," CJ said. "This is your place of knowing, your clairvoyance. What animal do you see there?"

Just then, a fly landed on my forehead and walked around. I swished him away; he landed again, and then again. "Okay!" I thought. "Fly — Garbage-Head, the fly — clean up my thoughts. Whenever you come, I will notice my negative thinking. Then you can eat the mental garbage and fly away."

For the seventh chakra at the top of my head, I saw the bat who flies to the sky. For the eighth chakra, I chose two power animals: the giraffe that reaches to the sky and the white spirit buffalo that runs in the heavens.

I slept soundly that night, but I was awakened early the next morning by CJ pounding on a drum. She did everything she could to hurry me, which I resented. As she hastened me to pack my backpack, I gathered a sun cap, sunglasses, sunscreen, and a full water bottle. I laced my hiking boots and was off.

I felt no fear until that moment. So I made up a song to sing as I hiked.

Hey now, everybody
Hey now, sing along with me
Hey now, everybody
Joy and light and harmony

I was determined to have a positive experience.

Hey now, animal kingdom
Hey now, fish of the sea
Hey now, birds of the air

Hey now, humanity
Sing now, everybody
Sing now, in harmony
Sing now, everybody
Joy and light and harmony
Joy and light and harmony

I hiked along the stream for about a mile to the place I had chosen the day before and made camp, setting up my tarp and sleeping bag. I was in a grove of trees on a bank with a steep slope above a sandy beach, where I could see and hear the water trickle down the brook. The ground was cushioned with pine needles. Mother Piñon stood at the east, Father Cedar to the west, and on the north and south were Brother and Sister Oak. Nearby was an abandoned homestead with the ruins of a hundred-year-old log cabin. Giant sagebrush stood amid rusty horse-drawn farm implements. Six- to eight-foot-tall sage bushes covered the field; their fragrance permeated the air, like incense constantly burning or the aroma of smudging sticks. My home was cozy, peaceful, fragrant, and shady.

Hey now, Mother Earth
Hey now, Father Sky
Hey now, insect nation
Hey now, plant kingdom
Sing now, everybody
Sing now, in harmony
Sing now, everybody
Joy and light and harmony
Joy and light and harmony

I filtered water from the stream into a jar to make "detox" sun tea. I was supposed to fast for three days, and from previous fasts I had done, I knew herbal teas eased the nausea. Neither the presence of bears and mountain lions nor the act of fasting frightened me. Bears, like dogs, scavenge for food, and I had none — except an energy bar in case of emergency. Mountain lions have a natural fear of humans and usually avoid us. What I feared was my inner world, my demons, the hateful thoughts and self-hatred — my inner bears and lions. I had no hope of avoiding myself now.

I prayed for gentle lessons, promised to pay attention to the messages, and asked for a happy experience. The first part of the day felt free and easy, with meditation, tai qi exercises, and short hikes. Each time a fly landed on me, I examined my thoughts. Sure enough, they were negative; the fly was a gentle reminder that left as soon as I found a positive cerebration in mind.

In the afternoon heat, I meditated in the loud silence of nature. With no distractions, I reached an inner clarity more profound than I had known before.

When I opened my eyes, I saw something on the piñon branch that extended under my tarp roof: four tiny orange caterpillars with yellow heads. The adorable little larvae had smiles on their faces as they did acrobatics to move from one needle to the next. They dangled upside down, curled, then pulled up, and stretched down or sideways to grab onto another stem. Their flexibility amazed and entertained me for a long time. I looked for more in the rest of the tree but did not see any.

Because there were only four, I accepted the caterpillar as my spirit animal, my vision, since Iron Feathers said the spirit guide would appear four times. The caterpillars symbolized my transformation from a worm to a butterfly or the growth of my soul out of its childhood.

That night, alone in the dark with no light and nothing to eat, I surrendered. Entertaining no thoughts of fear, I was calm. In the silence, I connected with my inner guidance. It was always positive, always encouraging and gentle.

The next day began much the same. In the heat of the day, I returned to my tarp, and watched the caterpillars for a while. Then, I reached for my journal in a side pocket of the daypack, and there, on the top edge of the note pad, was a small, white, fuzzy caterpillar.

"Hello," I said. "Who are you?"

"I am Caterpillar, and so are you," she said.

I picked up the journal and raised her closer. She was beautiful and seemed to be looking right at me. I felt honored by her presence but was unsure what to do with her. Concerned that I might accidentally injure her, I tossed her off the journal onto the ground, at the side of the tarp. She seemed offended; she pointed her nose to the air and walked off with amazing grace and strength, climbing up, over, and around every obstacle with ease. I wanted to call her back — to talk to her — but instead, I marveled at her agility and watched her disappear into the brush.

Many more visitors arrived. A large, black bug dug a hole shaped like a volcano next to the tarp. He left tracks all around the hole, four and four — an arachnid. Only the head was visible in the hole, until an unsuspecting fly happened by. The spider grabbed the fly and retreated in the blink of an eye. One less negative thought in my world was alright with me. A hummingbird landed on my tarp string and preened himself. He told me to go bathe.

The stream was shallow, so I walked along searching for a pool deep enough to sit in. I thought I had found one when a snake lifted his head from it and said, "Oh, no, you don't." I took a shower standing on a sand bar using a water jug.

The second night brought difficult dreams, and I woke into complete blackness, feeling nauseous and smelling a strong odor of skunk. The stench and a torment of mental images turned my stomach. I tossed and turned.

What does skunk mean? I asked myself. *Self-respect, skunk has a reputation that precedes her. She is gentle, but others respect her.*

At that moment, I realized my need for self-respect, and to my amazement and delight, the odor completely disappeared.

On the third day, I was supposed to build a "death lodge," an area to remain in from sunset to sunrise, where I would die and be reborn. Since I was a caterpillar, I decided to build a cocoon or chrysalis where I would transform. In the sand below my tarp, I gathered rocks and branches and shaped an oval. I made a small fire pit in the center and stacked a pile of twigs in it. My provisions included a pad to sit on, a gallon of water freshly pumped from the stream, a rattle, a drum, and some warm clothing.

Afterward, I left CJ a sign at our designated message place about a half mile away. CJ left a bunch of flowers for me; I left a pretty rock. On my way back I hiked the stream, stepping from rock to rock, as I did every day. My inner guide cautioned me, *Slow down*, which I ignored. I slipped on a wet rock and fell, landing on my left wrist. I was shocked; I never fell. After ten years of karate and three years of tai qi, I was usually as agile as a caterpillar. The wrist pain scared me at first, but after connecting with the silence inside, I understood that all was well. I shook my wrist as instructed, and the pain went away.

In the deep canyon, as the sun set, I entered my cocoon knowing the night would be long. The only lights were the stars and those flashing on jets, which told me that the airports had reopened after the September 11 attack; the nation considered air travel safe again.

The cool, vast blackness allowed for spectacular viewing of stars. The sky reminded me of a planetarium. A huge sparkling star appeared in the east, bright, with long, multicolored rays. Its shape resembled that of a dragonfly, or a fairy, and I thought I was hallucinating because it appeared to dance around, changing from white to blue to green to yellow to red. Sometimes a red ball of light would flash at the end of a ray. Around the star a burgundy aura glowed. It looked like an angel or a Disney animation.

The night grew cold, and I built a small fire. A spider, like the one living in the volcano near my tarp, ran quickly into my cocoon. He jumped onto a log in the fire and sat there next to the flames staring at me. The brave, bold spider was my reflection, a symbol of creativity. Together we discussed how to create the future before he ran off again into the darkness.

For a time, I sang and played the drum and rattles. I also embraced the stillness. Then, needing to move, I practiced tai qi arm swings to balance my chakras. I stood tall, keeping my centerline straight while my arms rotated around it. I focused from one chakra to the next, starting at the base, and acknowledged each time: *The first chakra is balanced. The second chakra is balanced.* Just as I thought, *The seventh chakra is balanced,* a bat swooped down low over my head. I screamed and collapsed on the ground. Then I remembered that the "bat who flies to the sky" was the animal I had chosen for my seventh chakra. Bat embraces the idea of shamanistic death and represents rebirth.

I sat alone in the silence the rest of the night, my mind still and clear as the cold mountain air. By dawn I felt elated and exhausted. As the sun peaked over the eastern ridge, I emerged from the cocoon.

Afterward, the hardest part of the entire trip was repacking the backpack. It had to be stuffed tight, and now I lacked the energy. Plus, I was hurrying to meet CJ by sunrise, as planned, and two girlfriends would be at base camp with CJ to welcome me back and hear my stories.

As I struggled with the tarp and the zipper, grunting and groaning and getting angry, a loud crash startled me. I turned to see two huge, brown eyes belonging to a bold but surprised bull elk. He had just crossed the stream. There he stood, staring at me. I thanked him, losing count of the points on his rack as he climbed up the slope into the woods.

Elk represents stamina. Elk medicine teaches us to pace ourselves. Another message elk brings is to seek the feedback and support of our own sex, according to the animal-based *Medicine Cards*. He brought the perfect message and a thrilling vision.

After my vision quest, I reflected on how veterinarians and shamans share several similar qualities and attributes. Veterinary school is the initiation where we are dismembered (or so it feels) and reborn in shamanic (veterinary) form. There we learn to communicate with animals. We become psychopomps escorting the souls of animals by way of death to their spirit home. We are healers and magicians who know that a turtle with swollen eyelids needs vitamin A. We can instantly and magically cure a paralyzed cow by giving her calcium intravenously. And as the shaman Iron Feathers mentioned, we struggle with the responsibility of the job, knowing that a normal life must be forsaken for our cause, to dwell in another realm with the animal spirit guides.

Perhaps death is like a lesson from the caterpillar. We enter a chrysalis, where our tissues are broken down, and then we emerge with beautiful wings and fly.

Ring of Fire

There have been three successive epochs in the history of the world.

In the beginning the animals were the masters, with mankind in a nebulous dream-state on the periphery. This epoch ended with the birth of the "transformer" or cultural hero, who rid the world of its primeval monsters and cleared the way for human life. All methods of hunting and fishing, the domestication of the dog and building of canoes can be traced back to him; it was through him that man became truly man. After his departure there began the third epoch, which still continues, and in which men and beasts can no longer communicate with each other, and only shamans maintain the link between the human and extra-human spheres. One day the cultural hero will return and then the earth will dissolve in flames.

— CALVIN MARTIN, *Keepers of the Game*[27]

Less than a year later, the teachings of shamanism helped me manage the difficult emotions around animal suffering and death during a massive forest fire in La Plata County. The winter of 2001 to 2002 saw no snow, our main source of water, and by June 2002, only 1.31 inches of precipitation had been measured for the year. Drought conditions were the worst in 107 years of recorded history. Daytime temperatures were in the high nineties, and the fields were yellow. Then on June 9, a fire started north of Durango on Missionary Ridge that sparked a conflagration, and by month's end it became a huge ring of fire, a hundred miles in diameter, burning over sixty thousand acres of wilderness lands.

The word *dry* took on new meaning when we learned that the live trees contained only half the moisture content of lumber. Some houses burned, but most were spared as the flames found the forest easier to consume. With each passing week, human emotions flared like the trees that exploded at the touch of an ember. The surviving wild animals were displaced and hungry. Hay and pasture for domestic livestock were scarce, as was the irrigation water needed to grow grass. Everything felt ready to dissolve into flames.

I was blessed. I had free use of large irrigated fields near the town of Ignacio, where my horses lived and grazed. My friends Earl and Janet Jones owned the property and treated me like a daughter. Earl was eighty-five, and Janet ninety-one. Janet's parents had homesteaded the ranch over a hundred years ago, and the water rights they possessed provided enough irrigation water to keep the fields alive and my horses well fed. Earl decided he could no longer work, and he asked my husband and me to irrigate the front half of his property. The other half of the sixty-three acres was leased by the neighbors, the Schmidt family, for cattle grazing.

Later that month, as the fire raged, I met Dolores Schmidt at Earl's front gate early one morning. She was a hardworking rancher with the shoulders of a college football player. "The prairie dogs sure are bad this year," I said.

"Oh, I tell you, they really take over in these dry years," Dolores agreed.

"The whole field is teaming with them running in and out of holes. It looks like a wound full of maggots."

"I like to let the irrigation water fill their burrows, and when they come out, I chop their heads off with a shovel," said Dolores. "The other day a whole family came out of one hole. There was a mother, a daddy, and three babies. They just kept coming. People think they're cute, but they wouldn't think so if they saw what they do to a field."

I drove into town shaking my head, feeling confused about prairie dogs and what Dolores had said. Ranchers tend to battle with the forces of nature, such as weeds, weather, and "pests" like prairie dogs that destroy pasture grass for livestock. Yes, prairie dogs are cute and have a right to life, but the same is true of horses and cattle that compete for the same food. I even think maggots are cute and important — little baby flies. Everything has a right to life, and we find ourselves trying to draw a line somewhere. We kill mice in our kitchens, mosquitoes in our bedrooms, internal and external parasites in our domestic animals — worms, fleas, ticks, lice, bacteria, and viruses. Paradise was lost; now we kill for food and to protect ourselves and our animals.

Dolores's callous treatment of prairie dogs certainly offends some people, just as animal rights activists offend ranchers. I can understand both points of view, and I reconcile the matter by trusting the spiritual teachings that say death is an illusion. First of all, prairie dogs are immortal! At least they seem that way when one tries to eradicate them from a field. I hope the teachings of shamanism are true, that there is no death, only transformation into the other realm.

Or am I just fooling myself in order to justify all the euthanasia I have performed and to soothe my sorrows after seeing so much death? I, too, long for a paradise where humans live in peace with nature. I imagine a world, the invisible world from this perspective, where everything is at peace forever.

During late June, with the fire raging, a foreboding sense of slaughter was in the air. Millions of creatures in the forest were dying of smoke inhalation, starvation, or incineration. As I drove through Animas, the air was so thick with smoke that cars had their lights on at nine in the morning. Through the haze I saw a semitrailer truck headed toward town carrying two red firetrucks on its bed.

Once again, my day as a veterinarian had begun with a call to kill — euthanasia was scheduled for a Dalmatian named Perdu. She was very old and paralyzed from the neck down. The owner, whom I had only spoken with on the phone, brought her dog home from the hospital the previous night and decided it was time to "put her to sleep." Upon entering the driveway, I saw three women in the woods sitting around the dog.

"This looks like a going away party," I said as I approached.

"How can you do this job?" asked a pretty, dark-haired woman.

I knelt down next to Perdu. "Well, I used to struggle with it, but now I feel more at peace being the reaper," I said. "Everything faces this transition. It's a special time, like birth." I paused, and the woman's face softened. "I used to have a hard time with this part of my job. Now, I feel honored to be with an animal at the time of its death."

The dark-haired woman smiled as she petted her dog. "I never thought about it like that before. I've really dreaded this moment."

"She's going on a glorious journey. Remember the ancient religious stories about paradise? At one time, we could travel between this world and the spirit world. We lived in peace with all of nature and could communicate with the animals. Then we left the garden and became stuck on earth, unable to cross over until death. But shamans retain the ability to go to the spirit world and talk with the animals. From them we know this is a joyous transition. Perdu will be set free from her suffering. I don't feel sad for the animals. It's the people they leave behind that miss them." Tears welled up in my eyes, and I blinked them away. "But she will always stay connected to you." Then I asked Perdu, "Are you ready to go home?"

"She is," said the woman. "She told me last night. Should I stay with her?"

"She would love to have you hold her as she passes over," I said.

An intravenous catheter remained in her hind limb from her hospital stay, so I injected through it. "Go quickly to the light, Perdu. Thank you for being such a good friend." As the spirit of Perdu left, the three women were still.

I drove into Durango and then headed north to visit a retired couple I knew well to vaccinate their two Morgan broodmares. There, the fire was visible on the cliffs above the Animas River valley to the east. Gay met me with a smile and a giggle.

"Well, we just have to grin and get through this," she said, as we both looked up at a mountainous mushroom of white and black fumes.

"The horses don't seem to mind all this smoke," I responded.

"No, they just keep on eating. Yvonne does look up at the cliff every once in a while, at all the smoke. Sometimes we can see flames. It's a little unnerving, but she doesn't seem too worried. We should be more like them, I guess. We're just glad to have irrigated pasture for a while, anyway."

"I heard that will end next week," said her husband, as he walked up leading the other mare.

Gay added, "A firefighter friend of mine was working up by the reservoir yesterday, and he said the wind was so hot birds were falling dead out of the sky."

Afterward, positive thoughts were elusive as I drove north. The fire had flared up into a heavily populated area the previous evening. To the east, the tops of the coral-colored cliffs stretched five hundred feet above the valley floor, and tall pines were bursting into ominous orange torches, sending plumes of soot shooting into the sky.

At the base of the cliffs were homes. If embers fell over the cliff, the valley floor would start to burn. Farther north, another fire was burning on the west side of the valley. Other drivers were pulling off the highway to witness this latest catastrophe, and over the radio I heard a notice for immediate evacuation of Animas Valley.

Just then, the cell phone rang with a call from Dawn — her dog, Apache, was not walking today. It had been a couple of months since I had last treated Apache. She had walked out of my office in late February 2002, seven months after her surgery. It took physical therapy to strengthen her hind limbs, and since then she had been doing well. But today Dawn was upset about her condition.

My drive to Dawn's office took me past the county fairgrounds, which had become the command center for the firefighters. Yellow, blue, and green tents belonging to hundreds of firefighters covered the grass in front of the adjoining high school. Hanging on the chain-link fence were handmade signs: "Thank You, Firefighters," "We Are Praying for You," "God Bless You, Firefighters." Passing cars also had signs on their windows that said, "We Love You, Firefighters." My emotions brought me to tears.

Dawn's office was in a windowless warehouse that she had designed to look like a campground inside. Dawn had created office spaces using canvas wall tents in a circle to resemble an outfitter hunting camp; each contained a desk, lights, and a computer. Between the tents lay patches of green indoor-outdoor carpet, with a picnic table in the center. Apache was asleep at the edge of Dawn's tent.

"I don't know, Karlene, she was doing great. We stopped the other day at the top of Coyote Creek Pass, and I let the dogs out. Before I knew it, she just took off. She went running down a steep slope chasing a squirrel, jumping over logs, and running around like crazy. Now she just won't walk. She dragged a patch of skin off her paw chasing a cat yesterday."

"Hi, Apache girl," I said.

Apache awoke and immediately stood up to come over to greet me.

"Look, just like when you take a car to the shop, it runs fine. She does seem weaker than usual, however," I offered, as I examined the pulses of her femoral arteries. "Let's jump-start your battery." It was cool and dark on the concrete floor of the warehouse, a pleasant escape from the heat and smoke. I treated Apache with electro-acupuncture and assured Dawn that all was well.

Back in the truck, the cell phone rang with another emergency in the town of Ignacio, near where Earl and Janet Jones lived. A laceration on a horse's nose needed suturing. As I drove, I passed a semi-truck carrying one-ton bales of hay. Help was arriving from every direction.

The owner of the horse was a longtime client and friend, a Baptist minister, Fred King. His daughter's show horse had a piece of skin hanging from one nostril.

"Well, if he has a modeling career, we better put him back together," I said, as I prepared my surgical tools. We discussed the community support during the fire. "I was talking to a shaman the other day, and I asked him why the natives don't have rain dances anymore. He said, 'Because they are all Christians.'"

Fred leaned back against the tailgate. "Years ago, a minister friend of mine lived among the Navajo. He lived there for many years and became trusted by them. One night, some shamans came and invited him to one of their ceremonies. He was very surprised to be included. They took him to a place where they made a fire with a hole in the center of it — a ring of fire. They took a chicken, cut its head off, and drained the blood into the hole. By morning, a six-foot tall corn stalk had grown there, with three ears of corn on it. He told me it was then that he realized what he was up against. That was satanic."

"Why do you say that?" I asked.

"Because there are only two powers on the earth, God's and Satan's, and with God's, you do not need to cut off a chicken's head."

Fred's answer helped me understand how the white man has influenced the native cultures spiritually. Before Christians and fur traders arrived in North America from Europe in the sixteenth and seventeenth centuries, native societies lived in harmony with nature. Then, when cod fishermen, fur traders, and European settlers came to this continent they brought with them colds, flu, streptococcus infections, typhoid, diphtheria, measles, chicken pox, tuberculosis, scarlet fever, whooping cough, yellow fever, smallpox, gonorrhea, and syphilis. With no immunity to these diseases, entire North American communities and populations were destroyed. Epidemics spread inland and devastated native peoples throughout the continent. In 1763, during the French and Indian War, the British gave Native Americans blankets covered with smallpox scabs, killing them by the hundreds of thousands.[28] Then, during the autumn and winter of 1781–82, a smallpox pandemic of appalling proportions ravaged tribes of the northern Plains and Upper Great Lakes.[29]

The shaman could not cure these diseases and ascribed the maladies to malevolent supernatural powers. The native tribes abandoned their spiritual practices when rituals failed. This corrupted their relationship with nature caused by the idea of retaliation and blame of animal spirits. The diseases not only decimated the populations but also broke tribal morale by crumbling the tribe's spiritual structure.

The fur trade increased the momentum of corruption between people and animals. The natives were starving, too weak to hunt. The fur traders gave peas, beans, flour, tobacco, liquor, and hunting tools in exchange for furs, thus opening the door to the unrestrained slaughter of fur animals. With commodities came the Christian religion, lending spiritual support to the idea that humans were superior to animals, and participation in the fur trade led to unyielding callousness toward nature. Some game animals were heavily exploited, while others were completely exterminated.

The native trader became locked into an alien economy, an unhealthy diet, alcohol consumption, a way of living with domestic conveniences, and a new religion that called their spiritual connection with animals evil.

I knew the Baptist minister had good intentions, but he did not understand the meaning of shamanism as I did. Blood spilled as the forest burned in a ring of fire, but in the center, where the blood is given, new life springs forth.

CHAPTER 4

Mother Nature: The Great Transformer

The Great Spirit is in all things, he is in the air we breathe.
The Great Spirit is our Father, but the earth is our mother.
She nourishes us; that which we put into the ground she returns to us.
— BIG THUNDER (BEDAGI) WABANAKI ALGONQUIN

As prehistoric humans shifted from a wholly nomadic lifestyle, pursuing animals and foraging for edible vegetation, to an increasingly agricultural one, cultivating plants and domesticating animals, they settled into village communities year-round. Hunter-gatherers gradually became farmers and herders. As this happened, shamanism did not die out; rather it blended into future religions. Like the tribal nomads, the first agriculturalists revered a unity of the Great Spirit and Mother Nature.

Early humans learned about life by observing. They watched, no doubt in wonder, as a bird hatched from an egg or a butterfly emerged from a chrysalis; perhaps amazed, they discovered that bees made honey, and seeds sprouted and grew into trees. Primeval humans believed Mother Nature was a magical spiritual being, as some still perceive her to be today.

Perhaps out of nostalgia for paradise, prehistoric humans protected ancient groves and left them undisturbed as sacred places, according to historian and archaeologist W. R. Halliday.[1] In these groves, sacred animals like white horses grazed, and tame wildlife such as deer mixed with wolves and allowed humans to stroke them. The springs and streams were holy, and fish were especially holy. Only their art remains to tell the story. These peoples may have been similar to the primeval Europeans who built Stonehenge, and the first Druid, Norse, and Greco-Roman peoples. The spiritual beliefs of these pre-Christian, agricultural clans are now called "pagan."

The religious awe of nature was the essence of pagan beliefs, and the agricultural lifestyle developed together with the pagan religions. Ken Dowden writes in *European Paganism*, "There is no beginning of paganism. It has no founder, no holy book that, once written, defined it. Before Christianity by definition all societies were pagan. The only alternative would have been atheism, a lack of religion all together. It is an interesting fact of human history that there is no evidence of atheist society, something which encourages the view that religion is in origin a dimension of society."[2]

Today the word *pagan* implies "nonreligious" or "polytheistic"; however, the original definition is "rural person." The anti-Christian connotation of paganism comes from the first Christian emperor of Rome, Constantine, who sent out monks to spread Christianity. The monks reached cities first, and they converted people living in rural communities last. Those country folk were called pagan, which is Latin for *village*. Constantine forbade pagan practices like decorating trees and hanging things around animals' necks, which were crimes punishable by seizure of house and land. For the early pagan agriculturalists, who had never heard of Christ, God was everywhere in nature.

Much of paganism is misunderstood, and many pagan practices persist today even within Christianity. We decorate Christmas trees and accessorize animal's necks as the ancient pagans did. The intention behind such acts is not to dishonor Christian beliefs. Rather, these practices existed prior to Christ and honor a different conception of spirituality.

The pagan religions were holistic. No chasm existed between God and nature. The divine dwelled in everything. Pagans did not worship the statues of calves any more than Christians worship the cross. Rather, they celebrated the life-giving "God Force Energy" in nature symbolized by the graven image. To this nature God, they also gave sacrifice.

Animal Sacrifice

In *A History of Religious Ideas*, Mircea Eliade explains that humans made a decision at the beginning of time to kill in order to live.[3] The gathering of fruits, roots, and mollusks did not provide adequate food for survival, so people began to hunt, as described in the Dreamtime story of the Australian Aborigines. The same problem has driven humans to hunt in more modern times.

As the first modern humans evolved, *Homo sapiens* developed a fellowship with the beasts. To hunt and kill a wild creature and later a domestic animal was equivalent to a "sacrifice."[4] For hunters and early agriculturalists, the act of killing an animal, whose blood appeared identical to their own, revealed a kinship between humans and animals. In the moment, the person felt connected to the wholeness of creation, and a mystical solidarity formed between the hunter and the hunted. These people believed that the spiritual aspect of nature took form in the flesh and offered itself in sacrifice as food, while the soul of the slain animal returned to its spirit home. They understood that the sacrificed animals offered themselves willingly.

Ancient pagan cultures venerated cattle as the most important animal. Cattle provided the source of all wealth, symbolizing fertility and abundance. The bull god, commonly mentioned across pagan beliefs, appeared in the form of a bull on occasion. The Greek god Zeus is an example. The god was worshipped in animal form, which meant that in the case of the sacrificed bull, the god and the victim were identical; the god himself was sacrificed and the body eaten.

Animal sacrifice was a common practice among pagans. Sacrificial offerings of cattle, swine, goats, and birds were made to the god or goddess, whose names varied with location. Offerings honored the divine. They were made in gratitude, to ask for forgiveness of sins, and for something beneficial, like an abundant harvest. Only the inedible parts of the beasts were offered, such as the blood or the horns. The rest of the meat was for feasting.

Women cried out as the animal died. The murdering group shared and dispersed the guilt, united in communal camaraderie of praise for the meal.[5] They inspected the entrails of the sacrificed beast for omens and examined the zones of the liver (hepatoscopy), just as veterinarians evaluate carcasses today in slaughterhouses to protect consumers from diseased meat. A feast followed, with offerings to the divine. For the agrarian, as with the hunter, the sacrificial ritual was a way of dealing with the conflict and anxiety of killing and eating animals.[6]

Although today some believe we have transcended the concept of animal sacrifice, the practice of sacrifice in thanks persists today in the world's religious celebrations. In truth, our offerings of animals remain somewhat revalorized and camouflaged. In America, we harvest the older animals in the forest to keep the herds young and vital. We experiment on laboratory animals in the name of science. We sacrifice chickens, cattle, and pigs by the thousands for barbecue fund-raisers to buy swing sets for parks or to help pay medical expenses for cancer patients. We sacrifice coyotes because they kill our sheep, and we kill prairie dogs because they destroy our fields. We believe that Christ, "the Lamb of God," sacrificed himself for our sins, and we take the symbolic body and blood of Christ at communion.[7] Many believe that if we suffer and sacrifice ourselves, we earn entrance into heaven. Religious beliefs about self-sacrifice underlie the suicide bombings in the Middle East and the suicidal plane crashes in New York on September 11, 2001.

Animal sacrifice is an ancient and universal human expression pervading all religions. Aside from the religious component, animal sacrifice remains identical to the modern slaughter of animals for food. Slaughter and sacrifice are both supposed to be done respectfully and as quickly and painlessly as possible.

THE BEAR SACRIFICE

As an example of ancient practices, and of how people love the animals they sacrifice, consider the Ainu, an indigenous Japanese tribe. For them, the sacrificial animal was more than food; it was a holy deity that gave itself to feed the people. They practiced a ritual bear sacrifice that expressed gratitude toward and respect for the animal. They prayed for a fast death and for the bear's spirit to return to its spirit home.

The Ainu based their ritual on the idea that the bear was a visiting mountain god.[8] The deities visited earth in disguises and became locked in animal forms until they were released through sacrifice. The bruin offered its meat and pelt in gratitude to the people. The Ainu captured a bear cub, raised it with care, and at the time of sacrificial offering, celebrated a communion with their god by partaking of his flesh. The bear was sacrificed to himself and the worshippers.

During the ritual, they prayed: "O Divine One, you were sent into this world for us to hunt. Precious little divinity, we adore you; hear our prayer. We have nourished and brought you up with care and trouble because we love you so. And now that you have grown up, we are about to send you back to your father and mother. When you come to them, please speak well of us and tell them how kind we have been. Please come to us again, and we shall again do you the honor of a sacrifice."[9]

Raising Livestock and Modern Industry

The respect and gratitude ancient peoples showed to sacrificed animals is shared by many rural meat producers today. In later chapters, I examine further the spiritual implications of killing animals and eating meat, but here I share some personal insights from veterinary medicine and stories of people who manage livestock in La Plata County.

In school, I took meat science, which was a required course for a pre-veterinary major. In this class, I learned how to take care of animals to produce the best meat. Reducing stress for animals is important; if animals become stressed prior to slaughter, the hormones released negatively affect the meat. Cruelty exists with animals just as it does with humans. However, a built-in karma exists with animal husbandry. Experienced livestock producers know that what is good for the animal is also good for business. Animals who are well cared for reward their caregivers with good production. On the other hand, if animals suffer from overcrowding and stress, heavy parasite loads, or poor nutrition, they do not make as much milk and meat. They don't reproduce as well; they become infertile, ill, and die, all of which cost money and are bad for business. Good producers know that even in transport one must not injure an animal because bruising ruins the meat.

Our class took a field trip to a cold-cut production company. That was enough for me to swear off wieners and bologna and choose a vegetarian diet for most of fifteen years, until health concerns drove me back to eating meat. I still avoid cold cuts; the ingredients used in these products disturbed me most. This experience opened my eyes to what goes into the food we eat and how important it is to read labels. Another important thing I learned about food is how the consumer drives the market. With our purchasing choices, we change the world.

How wonderful it would be for everyone to return to paradise, live in harmony with nature, and eat only fruit. Yet the reality remains that billions of people eat meat (and use meat in pet food), and the livestock industry thrives across the globe. In the United States in 2007, livestock sales accounted for about 52 percent of the nation's total market value of products.[10] The industry covers all fifty states and employs millions of people. It produces around $15 billion in income taxes annually and another $6 billion in property taxes.[11] Since this huge industry will not soon disappear, the most important matter for veterinarians, producers, and consumers is the humane care of the animals.

The US Department of Agriculture (USDA), together with the American Veterinary Medical Association (AVMA), enforces laws regarding the humane transport and slaughter of animals. As a member of the AVMA and an accredited veterinarian working on behalf of the USDA, I promise that we take these matters seriously. For example, the AVMA "Animal Welfare Principles" state that, among other things, "Animals shall be treated with respect and dignity throughout their lives, and when necessary, provided a humane death."[12] Many misconceptions about the livestock industry contribute to criticism of those who manage animals for food. Concerns are warranted, of course. However, most people do not understand how animals are raised or how meat is produced.

Let us look at beef. Similar to the ancient pagans who valued cattle the most, beef cattle production represents the largest single segment of American agriculture.[13] Critics claim that most cattle are raised in "factory farms," which is a derogatory term for a concentrated animal-feeding operation, such as feedlots. However, the majority of cattle are not owned by large, heartless corporations, and they do not live in feedlots. According to a March 17, 2015, USDA report, 97 percent of cow-calf operations are small family businesses (687,540 farms) that own an average of forty head of cattle. The majority graze on pastures where crop production is not feasible. They live in herds comprised of cows and their calves. Cows are kept and bred to produce calves, some of which are sold to other breeders, who raise the cows and bulls on pasture. Other calves are raised on grass until they go to slaughter. Only about 40 percent of the cattle go to a feedlot for an average of 140 days and are fed intensively, or "finished," with grain (mostly corn and soy) to fatten them before slaughter. Feedlots are designed to increase the amount of meat each animal produces as quickly as possible and to produce prime cuts of meat full of fatty marbling.[14]

Personally, my joints, muscles, and connective tissues become inflamed from corn, soy, wheat, and other grains. This miserable pain has led me to avoid most grains and feedlot beef. The amount of grain-free foods, and pet foods, available today tells me that many other people have similar ailments to mine. The demands of the consumer determine what stores sell, what products are made, and what type of meat is produced.

It is important to understand that each purchase a person makes creates a demand for what stores will sell in the future. For example, in recent years, more people have been buying organic foods. In the United States, this has led to three out of four conventional grocery stores joining the twenty thousand natural food stores that now sell organic food.[15] This trend exists on a global scale.[16] Here is how our purchasing power works to help animals. When my favorite brand of yogurt was no longer available at the health food store, a clerk told me it was because that company did not treat their cattle humanely. I am grateful that a retailer cares to keep track of such matters. I choose to shop at such stores. Feedlot meat can be avoided by purchasing grass-fed and grass-finished beef. These sales support the small family farms where cattle live happily on pastures.

How can people who love animals kill them and eat them? Food is a good reason. Livelihood is another good reason. Ranchers feed their families and make a living raising animals for food. Ranchers share an agreement with the beasts; the people feed the animal for a year or two, and the animal feeds the people for a year or two. Here, in the semi-arid desert and mountains of the southwestern United States, cattle graze where vegetables do not grow. One client described her childhood on a New Mexico ranch: "It was a hard life, but a good life. Death was fast for the animal. We knew what went into our meat, and we used every bit of it." These families have to care about the animals in order for both to survive, and it is not an easy job. Still, they enjoy it, and the critters provide endless entertainment.

As examples, here are two personal stories about people who raise livestock.

A Cow Named "You"

A friend of mine, named Gritty, made it a policy to never name his cattle. He felt it was best not to become attached to an animal destined for slaughter. But one cow earned the name "You," as in "Darn, *You*," "Dang, *You*," and the like. She was a mean black Brangus, which made her a good mother. When her first calf was missing, Gritty set out in search and noticed that You had been standing along the fence by the woods. There, on the wrong side of the fence and hidden in the tall grass, lay the calf. When Gritty's son picked up the calf to bring her back to the ranch, You jumped straight up from a standstill and right over the five-strand barbed-wire fence and charged the boy, who dropped the calf and hid in the juniper trees until You calmed down. You earned her keep by producing a good strong calf each year. She also caused Gritty much frustration with her habit of jumping the fences in search of more preferable food. One day, You was missing. Gritty knew right where she was. He got in his truck and drove the five-mile trip down his long driveway, along the county road to his neighbor's driveway and up to their barn, where he found You eating from his neighbor's hay stack. Gritty got out of his truck and yelled, "Goddamn, You, go home!" Gritty then got back into his pickup and drove to his ranch.

People might think cattle are dumb, but You understood who Gritty was and what he meant. She knew her name, and home she went. When Gritty arrived back at his place, You met him at the gate. With her ability to jump and charge, You could have gone anywhere she chose. Apparently, she preferred to live with Gritty, in spite of his cursing. Some may believe that cattlemen are brutes, but that's not true. They spend their lives looking after these beasts, and the cattle look after them, too.

Clearly, animals make choices. The question I ponder is whether, from the spiritual perspective, they choose to offer themselves as food as the ancient shamans and pagans believed they did, or if this idea is a human creation that soothes our guilt for killing them. I will examine this question further in later chapters.

In any case, anyone who enjoys butter on their potatoes, or wears a leather belt, participates in the lives and deaths of animals. Every person who eats meat, dairy, and eggs, uses leather products, or buys pet food benefits from the livestock industry. Rather than passing moral judgment on those who raise meat, I feel the most appropriate sentiment is gratitude. I am grateful to all the beings that go into the creation of my food. The people who take responsibility for the lives and deaths of food animals are not typically cruel; they do not enjoy the kill. Most love the animals, or they would not be able to handle the challenges of raising them. Rather than judging, I choose to take responsibility by making wise purchasing choices to support the people who care about livestock, such as the local ranchers who live in the communities of the southwestern US.

THE FARE OF THE FAIR

Children raised with farm animals learn early about the cycle of life, just as the ancient pagan agriculturalists did. One can find such children, particularly members of the 4-H Club, proudly showing their animals at county fairs. I have long supported 4-H Clubs because they teach children how to care for and be responsible for animals. The *Durango Herald* reported a story of one such child, a nine-year-old girl, who took her sow to compete at the local fair. The girl had worked diligently: She kept detailed record books, studied feeds, bathed the sow, and attended clinics on how to handle the animal in the show ring. The two-hundred-plus-pound pig far outweighed the girl, yet it obeyed the taps of the whip telling it where to place each limb. Together, they won the blue ribbon.

On the last day of the fair, as always, the animals go up for auction. The local businesses support the children by bidding on the animals and purchasing them for meat. The little girl's prize-winning sow sold for one thousand dollars.

A local newspaper reporter interviewed the child, asking how she felt. She frowned and said she felt sad about the sow being slaughtered. When asked what she would do with the money, she smiled again and said, "I'm going to buy pigs and feed them."

The child enjoyed caring for pigs. She polished the sow for show and hung a ribbon around its neck, yet she offered it up for sale for meat just as pagans worshipped their animals and sacrificed them for food, since that was why they raised them.

The Sacrificial Fire

In 2002, the Missionary Ridge fire that raged through June and into July was the worst in La Plata County history. Many animals died, and the fire created its own weather system. Winds fanned the flames and caused tall pillars of smoke to rise in the distance beyond the town, and daytime temperatures reached the nineties. Each night a weather inversion occurred, and smoke settled in the lower elevations. I had to close my windows, or the house would fill with smoke. That, along with my menopausal hot flashes, meant I had trouble sleeping.

Each morning the smoke was so thick that I could barely see horses standing in the fields, yet they never seemed concerned or upset. Each day on the way home from work, I visited my neighbors, Stacey and Joel, to take fresh-cut grass to my horse, Sport. Sporty, my white, seven-year-old, quarter horse–Arabian cross, was stalled there healing from a broken pelvis. Stacey and Joel had kindly provided a pen for him at their house.

Sport had broken his pelvis three months before, and I did not know how. No one witnessed the accident. I simply found him standing next to a large wooden fence post, which was broken off at the ground. Perhaps he spooked at something real (like a mountain lion) or imagined, reeled backward, and collided with the post. Sport crushed the right ischium and separated the pubis, causing the two sides to override each other. The right ilium must have also cracked, and both sacroiliac joints were dislocated.

Many people told me Sport would never heal, including the veterinarians at Animas Animal Hospital. I ignored their opinions. Pelvic fractures are not uncommon in horses, and they often mend with rest. Admittedly, Sport's injury was severe, but I hoped he would heal well with chiropractic, acupuncture, added nutrition, and confinement. I fed Sport Chinese herbal formulas for pain, to aid in bone mending, and to calm his disturbed state of mind. I adjusted his pelvis with weekly chiropractic to align the ischia, ilium, and sacrum. And I performed acupuncture to relieve pain and increase the circulation of blood to the area. Fractured bones must stay still to grow together, so I confined Sport to a small pen at Stacey's for three months. A severe fracture requires extra energy to heal, so I fed him abundant amounts of hay, grain, fresh dandelions, and mineral supplements. He had recovered enough to lie down, roll over, get up, and trot, and I planned to turn him out to pasture the next day.

As I drove up to the property, I saw Stacey and Joel. They stood in the yard with binoculars watching the firefighting planes and helicopters dropping water and slurry around the communications towers atop a nearby mountain.

"I can still see the towers," said Joel, looking through binoculars. "But it doesn't look good. Those columns of smoke are huge. Here comes the sky crane."

"How big is that bucket of water it carries?" I asked.

"About the size of a twelve-foot-round swimming pool, maybe holding a couple thousand gallons," replied Joel.

"It looks like spit next to that bank of smoke," I added. "The radio report said they made a four-bulldozer-wide track for a fire line around the communication towers."

Joel's face showed his worry. "Well, the fire has jumped a lot of fire lines, and it really wants to run up the top of that ridge. If it gets through...a lot of people live up there." He paused, voice weakening. "I think I'll wait to cry until this is all over."

I left my friends at dusk, the time of day when the fire always seemed to get worse. Then my cell phone rang with another emergency. This time, an injured llama west of town needed help, twenty-five miles away. A celebrity interested in conserving wildlife owned the ranch but only lived there part-time. In the first one hundred yards of the driveway, I saw three cow elk, five mule deer, and two beaver along the running stream. I drove past the houses to the pasture to meet Robby, who had been given permission by the ranch manager to move his llamas there to get out of the path of the fire. Robby stood in a field with three llamas; one of them, named Red, was injured. Robby held his two-year-old daughter's hand as he talked on a cell phone.

"I need as many horse trailers as you can get here now," he said. Then he looked at me, "Karlene, I need to show you something." He smiled at his daughter lovingly. All three of us left the llamas and walked southeast through the pasture until we came to several large piles of bear scat.

"There's nothing in this scat but grass," I said. "The bears must be really hungry."

"They are," he said. "Follow that trail of wool. I'll stay here with her." He smiled again at the happy little girl.

I followed pieces of white and brown llama fiber through a cottonwood grove to find a dead llama caught under a gate. The hind end was completely eaten, and one hind limb was missing. I walked back to Robby. "I never thought a bear would take out a four-hundred-pound llama."

"He was my toughest male," said Robby.

Knowing the territorial behavior of male llamas, I said, "I bet he was defending the others."

We walked back to examine Red, the injured llama. His breath was rapid, and he groaned and grunted as he got up and down, unable to walk. Three of his fetlocks were obviously dislocated or worse. We didn't know what happened, but my guess was that he ran to avoid the bears, and he either collided with something or landed badly in a depression on the ground, which strained or tore the ligaments in his joints.

I said, "I can't take X-rays out here without electricity. I'll give him some pain medication and wrap the limbs. We can take the pictures later at your house."

Then, a piercing, squeaky, turkey-gobble-like warning call came out of a brown male llama behind us. We turned in time to see a large black bear run across the meadow on the other side of the cottonwood trees. "There he is, right there; he isn't more than sixty yards away," I said.

Robby's wife drove up in a Suburban and came over to pick up their daughter. Robby phoned his brother. "Get your three-thirty-eight and come over here now. A black bear killed a llama, and he is coming back for more."

"There's another bear," I shouted. "It's smaller, and it's coming to the carcass."

"Bring my pistol," Robby said to his wife. "I may need to defend the other llamas."

The brown llama screamed another warning. His eyes were fixed on the area in the cottonwoods where his dead companion lay under the gate. Red, the injured llama, whined in distress, getting up and down, and panting as I tried to examine him.

"He has a fever of a hundred and five," I said. "He may have a wound under his wool somewhere, or the fever may be from stress and pain." I gave him Banamine for inflammation and penicillin for infection. I rubbed topical analgesic over his fetlocks and wrapped them for support.

Robby had remained composed while his daughter was with him, but now the stress of the situation showed. "I have kept my llamas here for four years without trouble. I have thirteen here now that I have to move tonight."

It was almost dark when Robby's brother, Jim, arrived, gun loaded. He looked through a large pair of binoculars at the hillside and said, "There he is, at the edge of the woods."

"That's the other one," said Robby. "There's one right here in those trees eating my llama."

Jim grumbled, "The fire jumped the east road today. Both the Bar K and Plotski's are burned. Someone is going to get sued." Jim stomped out into the field. "There's been a fire on that ridge for a week, and no one has done a damn thing about it."

I was confused. "Who are they going to sue, some volunteer who's risking his life for property?" I asked. "There are thirty major fires in the country right now and only so much manpower available." Jim did not hear; he was halfway to the bear.

"He gets excited," said Robby.

Jim climbed over a fence behind the carcass. Then, *bam*, his gun fired. The bear started to bawl. "He got him," I said. The llama let out another warning gobble as the bear limped up the slope in the oak brush. Jim marched after him, but the wrong way. We called out and pointed, and Jim moved to the left, then, *bam*. The bear ran out into the meadow, as the llama squealed another warning call. *Bam*. The brown llama stared intently as the collapsed bear continued to lift his head. *Bam*. Jim marched up the ridge in search of the other bear.

Robby needed to get back to the business of moving his herd. He said, "I had to evacuate the llamas and bring them here because of the fire. But it looks like it's safer to take them back home. Can you come and take X-rays tomor-row? This is my best packer. I'm sorry you had to see all that." He shook his head.

"Oh, I'm okay. Those bears are better off with a quick death than starving. I'll come by tomorrow."

I headed down the mountain toward Durango in the dark of night, and from the top of the hill at Hesperus I saw the east side of the valley glowing with flames. My emotions were all over the place. I needed to talk to someone. My husband, Jean-Luc, was in Utah helping to lead a vision quest, so I called Earl Jones, the rancher who owns the property where I pasture my horses.

"Did I wake you?" I asked

"Oh, no, we're just listening to the scanner."

I told Earl about the llamas and the bears and the view I was seeing. "We're recycling everything, bears, llamas, and every tree in La Plata County."

Earl agreed. "Yep, we must have done something wrong for God to punish us like this. We drove up the valley, after my chemotherapy today, to try to see the fire, but it was too smoky. We couldn't see anything."

"How are your treatments going?" I asked.

"Good. They say the tumor might go away completely."

"That's wonderful," I said. Then I complained about the drought, and Earl said, "Yep, it reminds me of the dust bowl days of the thirties. We planted po-tatoes, but they didn't grow." I asked what they ate, and he said, "We hunted." That put things in perspective. Trucks were delivering food from other places, so we still had plenty to eat. Sadly, the wildlife did not.

The next morning, I hooked up the horse trailer and picked up Sport from Joel and Stacey's paddock.

"Do you think he'll do alright?" asked Stacey.

"Oh yeah, he'll be fine"

"He moves kind of funny," Stacey said.

"Well, so does your dog, and he gets along okay."

"That's true. Do you think you'll be able to ride him again?"

"Absolutely. Healing happens. It's only been three months, and he can get up and down, roll over, buck, kick, run..."

Sport loaded into the trailer without trouble, and I took him to Earl's pasture. After being freed, he stopped at the first patch of green grass he found. I did some chores, then I walked out to where he grazed and called, "Come on, Sporty, let's go see your buddies." He followed of his own accord through the oak woods to the irrigation ditch, and then he walked into it, pawed and splashed a bit, got a drink, and headed up the steep bank on the other side. His hind limbs sprawled, he slipped, then he pushed up and calmly entered the pasture where the three molly mules (our "girls") and Pecos, my Appaloosa gelding, were grazing. Pecos walked up for a greeting. The three girls were more suspicious. Bonnie welcomed him with three quick kicks to the belly. "Bonnie!" I yelled. She stopped, looked at me, blinked, swished her tail, and went back to sniff Sport, who seemed unfazed by Bonnie's blows. He was too interested in eating grass to care what the mules did. They all settled in, and I headed off to work.

In the post office parking lot, I sat in the truck and read a letter from the woman whose dog, Perdu, I had euthanized only a few days before. The letter read:

> I wanted to thank you so much for guiding Perdu and me today. Your presence helped turn one of the worst days of my life into an unbelievable experience that neither she nor I will forget. My friend put it well when she said, "Leave it to Perdu to beckon a spirit like hers (meaning yours) at the time of her death." From the second you arrived, I knew I was in the presence of someone with a true gift. I am a Stanford biologist and Olympian and have met many remarkable people on my journey. And in a twenty-minute interaction, you are high on that list. In such a few words, you made sense of a situation that I truly have been dreading for years. And I will hear those words you spoke for the rest of my life....I really will. Thank you for arresting my fears and showing such compassion for *Homo sapiens* and canine alike.

There was no way to fight back the tears I had been saving for days. At that moment, the importance of companion animals and the words I use at the time of one's death struck me. The fire had burned fourteen more houses that night, and the air was so smoky it stung my nostrils. It filled the house at night if I opened any window, and the heat made it impossible to sleep. It was another day of temperatures in the nineties, and the winds were blowing strongly, feeding the fire. My emotions reached the breaking point. Tears flowed stubbornly at first, like rain after a drought, and just as rain relieves atmospheric tension, I felt relieved.

It was July 3, and the radio announced a gathering of the local Southern Ute tribe to dance and pray for rain at the south end of town at noon. The public was invited. Robby called to tell me that he had to meet with Division of Wildlife personnel to justify killing the bear. We agreed to take the X-rays at four o'clock.

Later, as I drove to Robby's, I noticed black clouds forming, then I saw drops of rain on my windshield. "Where is this coming from?" I said aloud. The weather channel had said, "No chance for rain." But it did rain. It was a gentle "lady rain," as the locals called it, a mild shower without thunder and lightning.

Robby lived in the ponderosa pines, and among the tall trees stood twelve llamas in the paddock. Red sat on his sternum (a position we call "cushed") in the yard next to the house. He could only walk a few feet before he cushed again. It took some time to get him under the deck where I had the X-ray machine set up out of the rain. Robby grabbed Red's wool and rolled him on his side. I took the pictures. The radiographs showed that the right front and both hind fetlocks were dislocated. I covered them in liniment, aligned them the best I could, wrapped them in bandages, and fitted them with splints that Robby helped to fashion out of PVC pipe. Red lay on his side for most of the process, something I had never seen a llama do without sedation.

"You see why I like him so much," Robby said. "He's my best packer. From the beginning he was easy to train. It doesn't look good, does it?"

"Time will tell. All I know is healing happens." I knelt next to Red, looking at Robby holding the llama's head. "My horse broke his pelvis three months ago," I explained to encourage Robby. "I mean, he really smashed it. Today, I turned him out to pasture, and he's doing fine."

"You're kidding! How can that heal?"

"Healing just happens; it's normal. You cut yourself and you heal. I don't know why people think things won't heal. I see animals recover all the time from supposedly fatal problems. Everyone said my horse wouldn't heal, but I didn't listen. It was difficult to look at his mangled rear end, so I kept looking at a photo of him that showed his round rump before the accident. I memorized it and visualized him looking like that again. You can't listen to negative people. Llamas really have it easy because they can be recumbent for a long time.

Horses have to be able to stand; they don't do well lying down for long periods of time, but Red can sit here cushed all day and eat. Let's feed him herbs for pain and change the splints every few days. Robby, picture Red on a high mountain trail with a pack on, moving normally. Let's see how he is in two months."

Red got up and walked to the paddock. The splints must have made it easier to walk because he only went down once on the way. "He's doing a lot better already," said Robby. "Ah, can I pay you later? With the fires, the insurance companies have put a stop to all new building in the county. No one can get insurance. I had some big jobs lined up, but now I don't know when I'll be working."

"No problem," I said. "At least it's starting to rain. Things will get better soon. So, how did it go with the Division of Wildlife?"

"They said there was no question the bear had to be killed. There was another bear eating on the carcass when we got there, so they shot him, too."

The policy of the US Department of Agriculture's Wildlife Services is to eliminate wildlife that kill livestock. The problem escalates once a wild animal finds domesticated animals easy prey, and they return time and again. This policy seems inhumane to the wildlife and has been questioned for decades. Still, it remains. These bears were starving. During that summer, with no fruit or nuts to be found, bears became nuisances all over. One entered a ninety-year-old woman's home in northern New Mexico and killed her. The claw scratches on the screens of Earl and Janet's home showed how desperate bears had become. Janet had poor eyesight, and I worried that if she heard a knock at the door, she might open it to a bear, thinking it was her big, hairy, neighbor Gilbert. The bear that scratched up the screens also destroyed some bee hives on Earl's property, even though they had an electric fence around them. The question is how to manage these displaced starving animals that destroy property and kill. Mountain communities have policies that work to avoid problems, such as a bear-proof trashcan rule. Still, bears entered people's homes during the drought in search of food.

I opened my mind to the possibility of a benevolent reality beyond this death and destruction and considered the view of ancient pagans. They described the world as a recycling of life. The death of an animal meant creation. Dogs and birds lived off scavenged bits of meat; blood fertilized the soil where seeds fell, and over time vegetation grew better there than elsewhere. Vines and grasses came out from between the bones of a carcass, and bees made a hive in the skull. Pagans worshipped the concept of fertility and the way that death brought life.

Pagans believed that death is like birth. This is the cycle of life. Both death and birth may be painful, but then with rebirth the love of the mother embraces the newborn, sharing her delicious milk. Science tells us that energy is neither created nor destroyed, so the energetic essence that leaves at death must go somewhere. Maybe the spirits of llamas and bears live on. The life-force of the llama went into the bear for a short time. The bear's meat provided life-force energy for ravens, crows, and other flesh-eating birds and small mammals. I started to look at roadkill carcasses differently; I prayed for the animals' spirits and thanked them for the bird food. All bad things have a good side to them. Fire is a natural process of renewal. The forests burned to fertilize the soil, open seed pods, and stimulate new shoots to sprout from the ashes. I consoled myself by trying to see the pagan point of view — Mother Nature is a giant recycling vat.

Gaia: The Goddess of Ancient Europe

She was the creator from whom all life — human, plant,
and animal — arose, and to whom, everything returned.
— Marija Gimbutas[17]

As I sought a deeper understanding of paganism, I studied the Mother Goddess religion of ancient Europe along with the religions of ancient Egypt and ancient China.

In ancient Europe, earth or Gaia — the Great Mother Goddess — represented a self-regulating organism, worshipped as the cycle of life, death, and rebirth. She was not divided into good and bad; she embodied that dualism as both creator and destroyer. The name varied with location: Inanna in ancient Sumeria, Ishtar in Babylon, Anat in Canaan, Isis in Egypt, and Aphrodite in Greece.

The ancient pagans conceived of their existence like a plant in nature. When one died, that person was sown back into the earth to be reborn like a sprouting seed. An animal's grave became a place where plants grew; a tomb was a womb, a sarcophagus a cocoon.

The Goddess religion relates to the Paleolithic and Neolithic ages, around 10,000 BCE, a time characterized by the discovery of agriculture, when sowing seeds for corn and wheat began, as well as taming grazing animals that provided milk, eggs, meat, and skins. The domestication of goats, sheep, pigs, dogs, and cattle also occurred in this era.

The art found on pottery, sculptures, and figurines of the Goddess and animals portrayed pagan life. Paintings depicted the important animals, which included serpent, crane, swan, goose, duck, owl, diver bird, vulture, butterfly, bee, dog, and deer. It included tiny creatures like the hedgehog, caterpillar, weasel, and toad as well as larger ones like the ram, pig, bull, cow, lion, bear, and stag. Masks of animal heads represented the intimate relationship of nature and humankind with the divine. When worn by a human figure, they embodied a fusion of animal and human forces. The mother animal with her child, whether bear, bird, or serpent, illustrated the image of the Great Mother's cycle of life.

An animal's transformative appearance remained significant. The bear symbolized death and regeneration. It went into its tomb or winter cave to hibernate, and it emerged reborn in the spring. The deer, whose antlers came and went seasonally, was an omen of renewal. Snakes symbolized regeneration with the periodic molt of their skin. The metamorphosis of the caterpillar into a chrysalis and then a butterfly showed how life changes from one form into another and demonstrated how souls survived and were reborn.

The humming of bees represented the voice of the Goddess. The earliest image of the Goddess is that of a bee carved into the stylized head of a bull. It is a very ancient belief that bees originated out of the dead carcass of a bull.

Cattle provided wealth in the form of meat, milk, and hides. The bull was the son of the Goddess. The horns of the bull corresponded to the uterine horns and the lunar crescent; they were thought of as communication devices, like antennae, associated with reception of godly knowledge.[18] The sacrifice of the male animal, the son of the Goddess, symbolized fertility, as new life came from the body of the bull.

Bird-goddess figurines (found in Crete, Greece, Egypt, Minoa, and Sumeria) were both lovely and fearsome, representing death and rebirth. Later, the Greco-Roman, German, Indo-European, and other patriarchal cultures vilified the images, such as those with serpents entwined around the body of the Goddess, referring to them as witches and monsters. Regardless of patriarchal or feminist opinion, remember that Gaia contained both the dual aspects of life. Sometimes she had doves or poppies on her head; sometimes she was entwined by snakes. She might appear resting in the shape of a bee or standing upon her mountain with lions, raising her arms as the wings of the bird goddess, or sitting beneath the Tree of Life offering the fruits to her priestesses.

Ancient Egypt

The Egyptians rarely (if ever) worshipped animals as gods,
but rather as *manifestations* of the gods. Animals functioned much as did cult statues, and were simply one vehicle through which the gods could make their will manifest, and through which the faithful could demonstrate their devotion to the gods.

— STEPHEN E. THOMPSON[19]

Ancient civilization in the Nile River Valley depended on the seasonal changes of the river. Each year snow melted in the highlands, bringing floodwaters to nourish the rich soil used for agriculture. As the waters dried up under the sun, the desert encroached upon the farmland. The fertile fields reflected the image of life, and the desert showed the lack of it. These contrasting forces of nature were the powers the Egyptians defined as divinities — the wet land and the sun.

The Egyptian creation stories describe the "First Place" as a pyramid-shaped hill of land that emerged from the primal waters. Another story tells of a primordial egg that contains Atum, the "Bird of Light," one of the main creator sun gods, who comes into being as raised land and light. In accordance with the symbolism of the serpent in other creation myths, the ancient Egyptians referred to Atum as the "primitive serpent." The Egyptian *Book of the Dead* prophesies that when the world returns to chaos, Atum will become the new serpent. Atum represents the supreme, hidden God, whereas Re, the Sun, is the manifest God. The ancient Egyptians worshipped many gods at different times and in different places over thousands of years.

The idea that the ancient Egyptians worshipped animals, historically a point of ridicule, is a misconception, according to some scholars. Others interpret the Egyptian hieroglyphics to indicate animal worship did occur. According to Professor Bob Brier, most Egyptologists agree that the ancient Egyptians did not worship animals as gods, with one exception — the Apis bull.[20]

The Apis bull was born of Hathor, a goddess often depicted as a human female with cow ears and a sun disc with a sacred asp between her horns. Hathor, represented by the great flood, nourished the world with her rain-milk.

Lightning from heaven impregnated the cow-goddess and brought forth the bull son — Apis.

The living Apis bull was the son and representative of the creator god Ptah. The spirit of the god passed from one ox incarnation to another, each one recognized by twenty-nine distinct markings. He had a diamond on his forehead, split tail hairs, and a scarab marking under his tongue, among other notables. The Apis bull lived in a palace and was pandered to, living as long as thirty years before he died, at which time the search for the next Apis would begin by looking for the markings.

When each Apis died, its soul joined Osiris, god of the afterlife, in heaven, and the bull's body was mummified. The Apis bull was the only animal mummified like a god. An ancient papyrus explained how the mummification was to proceed, describing every detail, such as how to wrap the god's leg. He was bejeweled and buried in a sacred place, the Serapeum (discovered in 1850), where an avenue of sphinxes led up to the tomb, the place where many Apis bulls were laid in sarcophagi weighing as much as five tons, with one-ton lids.[21]

The Egyptians believed in resurrection — they had tombs full of items to take into the next world, including animals and animal parts, mummified, as food for the deceased. Many other animals were mummified but for different reasons. The most common reason was as a sacrifice to a god. For example, if a person wanted to be healed, they would make a pilgrimage to the tomb of Imhotep, the god of healing, and make an offering of a mummified ibis. The ibis was sacred to Imhotep, and during the reign of the Ptolemaic kings, Egyptians sacrificed millions of ibis. A huge industry grew up around the production, mummification, and sale of ibis. Imhotep's tomb looked like a subway with piles of mummified ibis in pots stretching for miles.

People also mummified pets so they could travel to the next world with the family. The favorite pets of the Egyptians were the cat, the dog, and the baboon. They buried their dogs in sacred vaults amid great manifestations of grief. The Egyptians loved animals.

Sometimes gods appeared in animal form, as did Bastet, the cat goddess, but not all cats were gods. Myths about the Egyptian gods involved nature and agriculture. The god Osiris was reborn in grain. His sister and wife, Isis, the winged goddess, was an invisible principle of nature that manifested as various animals. Isis took the form of a kite, a scavenging predator with huge wings, to rejuvenate the dead by eating the rotting flesh and retrieving the soul. In other words, the Ka, or universal soul, appeared as a great bird hovering over the deceased, greeting the Ba, or personal soul, like a mother meeting her child, uniting the two into the breath of eternal life.

The same scenario played out with Anubis, the black jackal-headed god. The jackal ate rotting flesh by tearing the corpse to pieces and burying the bits for a time, digging them up later, thereby transforming putrefied meat into nutrition. The jackal portrayed the god Anubis, who guided the souls of the deceased and aided in the judgment of the dead.

Ancient China

The Chinese perception of the world did not insist on clear categorical
or ontological boundaries between animals, human beings, and other creatures
such as ghosts and spirits. The demarcation of the human and animal realms
was not perceived to be permanent or constant.... Instead animals were viewed
as part of an organic whole in which the mutual relationships among the species
were characterized as contingent, continuous, and interdependent.
— ROEL STERCKX[22]

Change is a concept central to ancient Chinese thought. The basic theory, oversimplified, is that all things emerged from a primal unity or undifferentiated field of energy called qi. From it, all the mineral, plant, and animal species developed through a process of metamorphosis, mutating from primal types. This theory maintained that the boundaries between the animated species were vague and indistinct.

A circle depicted the primordial essence or fundamental nature of the universe and represented the whole. A curved line through the circle divided the whole into two sides, one light and one dark, yin and yang, and duality was born — heaven and earth, male and female, and all creatures — with everything constantly changing, day into night, flowing on and on like a river. The focus was not on individual transitory things but on the underlying law at work in all change.

The *Book of Changes*, or *I Ching*, is the ancient book of Chinese wisdom, and it is unquestionably one of the most important books in the world's literature. Its origin goes back to mythical antiquity.[23] Although people today consider it a book of divination, it is much more. Both branches of Chinese philosophy,

Taoism and Confucianism, have common roots in the *I Ching*. Some modern scientists believe it is the manifestation of early science, with congruence between it and the genetic code, DNA, that, according to molecular biologist Gunther S. Stent, is "nothing short of amazing."[24]

The text originated in ancient times when a sage named Fuxi ruled all under heaven. He contemplated the images of heaven, the patterns of earth, and the markings of the birds and the beasts. From these markings Fuxi invented the eight trigrams (a collection of linear binary symbols) that later evolved into the *Book of Changes*.

These sacred writings appeared through animal mediums. One legend stated that Fuxi based the trigrams from the design on the back of a tortoise he observed emerging from the Luo River; thus, they were called the Luo Writing. Cambridge University professor Roel Sterckx offers this translation of an ancient text: "He looked upward and contemplated the round and crooked shape of the Kui constellation. He looked downward and examined the markings on the shells of tortoises, bird feathers, mountains and rivers, and then guided his palm to create the written character." Sterckx continues, "One source states that 'Fuxi's virtue penetrated the upper and lower realms. Heaven responded with patterns and designs of birds and beasts. Earth responded with the River Chart and the Luo Writing. (Fuxi) then standardized these images and composed the Changes."[25]

Animals also influenced the Chinese calligraphic style of writing, originally called "animal script" or "bird and animal script," because they were the "intermediaries transmitting or revealing sacred writing."[26] Bird traces and animal tracks and movements inspired calligraphy brush movements and hand postures. In order to understand Chinese calligraphy, one's eyes needed to be opened to the form and rhythm inherent in every animal's body and limbs.

The early Chinese did not analyze biological features in a scientific sense, but rather according to human moral analogy. The world of plants, birds, and beasts was a continuously changing physical reality, which included physical metamorphosis of animals and amorphous boundaries between living things. All animated beings started from a seed or germ. All minerals, plants, animals, and humans came forth from germs through a process of continuous mutation and transformation. A plant gave birth to a leopard, which begat a horse, which gave birth to man.[27]

Thus, they believed in human-animal transformations, with the transition of souls from human to beast and from beast to human. Animals changed from one form to another, which was seen as a sign of demonic power, good or evil. Metamorphosis and hybrid beings figured as regular occurrences that correlated to the yin-yang nature of the cosmic forces. Animal transformations were not seen as anomalous, but natural and indicative of the cosmic pattern of constant change. As Sterckx writes, "According to traditional calendars, hawks transform into pigeons in mid-spring, field mice transform into quails in late spring, small birds enter the big waters to transform into bivalves in late autumn, and fowl enter the big waters to change into mollusks in the beginning of winter."[28]

A modern scientific example of plant-to-animal transformation is the Cordyceps mushroom (*Cordyceps sinensis*), which is called "winter worm summer grass" because it appears to transform from an animal to a plant and back again with the season. Today there is still debate over how this fungus grows. Mycologist Malcolm Clark believes that Cordyceps spores are ingested by a caterpillar, where they germinate and then grow to fully occupy the caterpillar's body.

No major difference between humans and other living beasts was noted in ancient China. Humans, animals, birds, insects, demons, and spirits all possessed blood and qi, which gave them awareness. All these creatures, for example, were conscious of hunger and cold; they loved their own kind and would mourn their losses. Nonhuman creatures were observed to have impulses other than those vital to subsistence — instinctive temperaments and emotions linked humans and animals.

Blood and qi were more than physical properties; they contained the awareness and emotions. Species with qi and blood possessed the temperaments of joy and anger with the inclination to advance toward benefit and shun danger, thus preserving their species, in the same way as humans.

Since animals also contain blood and qi, regarding animal sacrifice, respect was to be shown toward all blood species. Sterckx writes, "Needless killing of animals amounted to a symbolic slaughter of the self and one's moral integrity."[29] Instructions for sacrificial killing required moderation. The act of bleeding the victim to drain the blood represented the process of transformation, changing the animal from an animate blood creature into meat.

The ancient concepts of Chinese medicine assign the body five spirits, each giving rise to five emotions, which both humans and animals possess: "The spirit of the heart is known as the *shen*, which rules mental and creative functions. The spirit of the liver, the *hun*, rules the nervous system and gives rise to extrasensory perception. The spirit of the spleen, of *yi*, rules logic or reasoning power.

The spirit of the lungs, or *po*, rules the animalistic instincts, physical strength, and stamina. The spirit of the kidneys, the *zhi*, rules the will, drive, ambition, and survival instinct."[30]

The *po* (the spirit of the lungs) is the corporeal soul, which is indissolubly attached to the body and goes down to earth with it at death; the *hun* (the spirit of the liver) is the ethereal soul, which leaves the body at death, carrying with it an appearance of physical form.[31]

Humans were not considered superior in an evolutionary sense, or the top of some hierarchy; humans were distinguished from animals based on moral principles, which differed more by degree than kind. For example, the animals provided the origin of music and dance in their sounds and movements, but only the human was able to master the raw sounds and turn them into music, which exerted a moral influence on all living beings, "enabling the sage-ruler to subsume the animal world under human control."[32] Humans were seen as superior in their sense of justice and social organization, or cultural skills.

Fantastic Animals

One of the common features of the early agrarian-age stories is the existence of fantastic animals like the dragon, unicorn, phoenix, and griffin, as well as other composite beasts that were part-human, such as mermaids, centaurs, and angels. The prominence of these fantastic beings in myth and art warrants further discussion.

Although these beasts are considered fantasy, with supposedly imagined magical properties, they are mentioned in books that are often translated literally, such as the Bible. The tortoise, considered a magical being in ancient times, is a real animal, and any number of dragon-like creatures also exist, causing me to ponder whether these fantastic creatures once lived among humans. A Tibetan Buddhist lama I know says dragons actually exist. Furthermore, non-dinosaur fossils from the Triassic period (about 200 million years ago) resemble dragons. We know that huge serpents fifty feet long, monstrous beasts, and noxious reptiles of all kinds once overran ancient Egypt.[33] Fossil evidence confirms that Australia's giant marsupials existed a million years ago, along with five-meter-long lizards, half-ton birds, and giant, dinosaur-like tortoises.[34] Some of the giant marsupials of Australia disappeared only fifteen thousand years ago.[35] Which creatures from ancient stories were real and which fantasies is beyond our knowing. Regardless of the current evidence to suggest that such beings may have been real at one time, their presence in ancient literature is so pervasive that, for me, they seem to have existed as literally as dinosaurs and pterodactyls.

Chinese writings from the Qin and Han dynasties (4000 BCE–1000 CE) offer intriguing notions about these fantastic animals. In ancient China, the four heavenly animals — the dragon, unicorn, phoenix, and tortoise — were present in the beginning and in times of harmony. The dragon or serpent is mentioned as a spiritual being capable of transformation. Confucius said, "The dragon is great indeed. While the dragon is able to change into a cloud, it is also able to change into a reptile and also able to change into a fish, a flying bird, or a slithery reptile. No matter how it wants to transform, that it does not lose its basic form is because it is the epitome of spiritual ability."[36] All creatures were described as descendants of the dragon.

The ancient Chinese believed that all animals emerged from an undifferentiated state and went through a dragon phase. According to Karen Armstrong, "In almost all cultures, the dragon symbolizes the latent, the unformed, and the undifferentiated."[37] In the myths of most cultures, a slaying of the dragon occurs. For example, in a Canaanite myth, Baal slays the seven-headed dragon, Lotan, called Leviathan in Hebrew (Isaiah 27:1). Zeus has to overcome the dragon Typhaon, monster with a hundred serpent heads, youngest son of Gaia, before he can establish himself as the father of all the gods. One scholar writes, "The Babylonian god Marduk fights a great battle to overcome the primordial salt waters of Tiâmat, who is portrayed in the account as the great Dragon Queen."[38] The theme of dragon slaying is thought to symbolize the conquest over ignorance or unconsciousness and the fear of chaos. The hero myth represents the story of the growth of consciousness.

Confucius referred to the dragon as the ultimate shape-shifter.[39] Perhaps he was referring to a creature that few have the ability to see today, or maybe he was referring to the serpent-like molecule DNA. Another idea of a shape-shifting dragon comes to mind from embryology. Indeed, embryology — the ability of an animal to develop from two cells — is one of the most spiritual, shape-shifting processes I have ever studied. All animals go through a somewhat "undifferentiated" phase in development that may be likened to a "dragon stage." Every being with a spinal cord develops from two cells (sperm and egg — yang and yin); then they form multiple-celled organisms and become embryos. The embryo, in the tail-bud form, is still somewhat undifferentiated. All embryos at this stage appear to be some sort of curled "tadpole." The embryos of fishes, amphibians, reptiles, and mammals have many similarities at this point.[40] From there they grow characteristics of their final determined species — fish, amphibian, reptile, bird, or mammal. The dragon, when seen as DNA or an embryo, is the epitome of transformative ability.

The unicorn is a symbol of purity. A number of Bible passages, in some translations, refer to the unicorn. For example, here is one translation of Deuteronomy 33:17: "His glory is like the firstlings of the bullocks, and his horns are like the horns of unicorns."[41] The early Chinese described the unicorn as a beast with the body of a deer, the tail of an ox, and a round horn in the center of its head.[42]

For the ancient Chinese, dying was changing into another human or creature. In the work of the Wang Chong, it states, "Living species transform according to their qi. One cannot deny this.... Sometimes in times of universal peace and when the qi is in harmony, a hornless river-deer may transform into a unicorn and a snow goose into a phoenix. This surely is the nature of their qi; they transform and change with the seasons."[43]

In harmonious times the most refined and pure qi produced the sacred animals of each species, and the spiritual essence of these beings appeared. The Da Dai Liji states, "The essence of the hairy animals is the unicorn. The essence of the feathered animals is the phoenix, the essence of the armored animals is the turtle and the essence of the scaly animals is the dragon. The essence of the naked animals is the Sage."[44]

The phoenix in ancient Egypt was said to have created itself from unformed matter. It arose out of fire, a symbol of resurrection. A person's spirit came out of his or her body just as the phoenix, the ka, sprang forth from the heart of the dead Osiris, the sun god. The phoenix was the symbol of the rising sun, a most holy being. Birds evolved from the phoenix in the following manner, according to the *Huainanzi*: "Winged Excellence gave birth to Flying Dragon. Flying Dragon gave birth to the phoenix. The phoenix gave birth to the simurgh, and the simurgh gave birth to ordinary birds."[45]

According to Chinese legend, the twelve-pitch standards of music were derived from the cries of the phoenix. The phoenix was also the originator of the wind instruments. Music was received and composed with the accompaniment of creatures like the phoenix and the approval of the spirit world.

In antiquity, the divine sages comprehended the cries of the animals and had meetings with them to receive instructions. As civilization degenerated, these animals became rarer. As man's virtuous conduct reversed, the creatures were frightened away, and the animals that previously had not caused any trouble began to attack people. The ancient Chinese observed this animal behavior to be a barometer of decline in human history. The sages stressed that the lavish hunting of animals would cause animals to chase humans. Laws were designed to keep hunting more respectful of the hunted beasts and the hunter's horses.

For the ancient Chinese, and other early agricultural societies, the observation of any strange, anomalous, or prodigious beings or any strange animal behavior was ominous. The Chinese admit that anomalies were not real but rather a creation of the observer, yet the observer's perception is real. "When people neglect constancy, prodigies arise. Therefore, prodigies exist."[46] For example, when two snakes were seen fighting at the gate to a capital city, it was taken as a prodigious event regarding the claim to the throne. All such events were explained as reflecting the qi of the observer: "When people fear something their qi flames up and takes hold of them. Prodigies come forth from man," according to the ancient Chinese historical work the *Zuozhuan zhu*.[47] In other words, freak animal behavior is attributed to irregularities in the human order. Similarly, shamans believed animals provided a reflection, implying that we create our own realities with our minds. For example, the qi overheats with anger and takes possession of a person, who then sees things from an irate perspective. When the person calms down, he or she may realize how the state of mind created a drama that did not exist. Sterckx shares the Chinese interpretation, explaining that the snakes were not real snakes but merely prodigious vapors that resembled snakes fighting each other.[48] Such vapors can arise from an individual as well as a group or even a country. The perception of freak animals and anomalous animal behavior reflect change in the affairs of human society.[49]

The conduct of the people determined the presence of the four sacred animals. As the corpus of the Warring States and Han writings indicate: "If you trample nests and smash eggs, the phoenix won't fly over the region."[50] The presence of the scared creatures changed with civilization. The capture of a white unicorn was an omen of the warring states of China overcoming barbarians, and of man's victory over bestiality, as well as the domestication of animals.

According to *The Yellow Emperor's Classic of Medicine*, people practiced the Tao, the Way of Life, prior to 350 BCE.[51] Humans lived in balance with the energies of the universe and nature. The ancient Taoists believed that achieved human beings could travel freely between time and space; sages were free of emotional extremes and lived in balance with the rhythms of the planets and the universe — the Tao. Just as with the creation stories of the Australian Aborigines (when humans left the Dreamtime) and of the Bible (when Adam and Eve were sent out of the garden), people lost balance with the harmony of the Tao. Our qi became obstructed, we began to perceive freak animal anomalies, and the pure forms — the dragon, the unicorn, the phoenix, and such — disappeared from sight.

Thus, to improve the world, we only need work on ourselves and cultivate a sage-like nature. Sages are not alarmed by strange events because they understand them. People only consider things strange when they do not see them regularly. The common person relegates unfamiliar creatures to the realm of ghosts and spirits, while the sage can differentiate image from reality because he or she has mastered "shape-shifting." According to the ancient Chinese, "Moreover there are no strange creatures of heaven and earth. When male and female mate, and yin and yang blend together, in the case of the feathered species it produces chicks and fledglings; in the case of the hairy species it produces foals and colts. The soft element becomes skin and flesh; the solid element becomes teeth and horns. People do not consider this strange....As for strange creatures under heaven, only the sage sees them. The ups and downs of their benefit and harm are only clearly understood by a knowledgeable person."[52]

The point is well made that things not commonly seen are given special significance by those unaware of them, whereas someone knowledgeable of such things finds them quite ordinary. For example, hysterical people telephone me to report some horrible, white ribbon-like thing hanging from their dog's anus. I am not alarmed because I know that it is only a tapeworm, which infested the dog when it ate raw deer flesh. Or people become upset because their horse's "teeth are falling out," until I explain that young horses lose their "baby teeth" just as people do; it is the normal course of nature.

Although I can read entrails and am familiar with tapeworms, I am not a sage. After all, I have not mastered shape-shifting. The ancient Chinese would no doubt agree, since animal doctors were low on their social ladder. One Chinese story mentions a horse doctor with a beggar, suggesting that the nobility scorned both. Nonetheless, horses were considered a valuable asset, and there was a state office for an animal doctor and horse sorcerer who probably practiced some shamanic and medicinal healing, as well as acupuncture.

One famous horse healer was Ma Shihuang, active at the time of the legendary Yellow Emperor (circa 350 BCE). According to one account, his healing skills were so efficient that a dragon spontaneously gave itself up to receive Ma's needle treatment.[53]

The tortoise symbolizes longevity and represents the beginning of creation. Perhaps the existence of the tortoise today indicates a favorable omen regarding the moral decline of humans and the nature of our qi. Perhaps if we each work to improve our own sage-like qualities, and find harmony with the Tao, our qi will become pure, and we may see the harmonious side of nature, and the essence of the spiritual beings will reappear. Maybe, on that day we rode in the Bisti Badlands, our qi was in such harmony that we were able to see a white owl, or maybe I only relegated it to the realm of spirit because I am not yet a sage.

Neo-paganism

Beyond the remnants of old-world paganism, neo-paganism continues to flourish. Paganism is common enough that in the United Kingdom's 2011 census respondents were allowed to describe themselves as pagan (druid, Wiccan, pantheist, shamanist, animist). Neo-pagans adhere to the religious practices of the ancient pagans, often described as "earth religion" or "nature religion." They hold a central ideology of the sanctity of Mother Earth and nature. People involved tend to be environmentalists, have feminist leanings, and may be animal rights activists. Some who identify with neo-paganism describe themselves as Wiccan, adherents of a modern pagan witchcraft practice with the creed, "If it harms none, do as you will."

The polytheistic deities of the "old religion," as it is sometimes called, are not taken to be literal entities but rather Jungian archetypes of the human psyche. Although some scholars argue that the pagans worshipped animals, many believe the ancient pagans actually held beliefs similar to neo-paganism and worshipped the aspect of God as expressed through various plant and animal forms.

The website Neo-paganism.com states that the neo-pagan theology is pantheistic; it views the material world as a manifestation of divinity. Pantheists believe God is in nature. Everything in the material universe, including animals and human beings, is a manifestation of God. No doctrine, creed, or requirements are included in pantheism. Even Albert Einstein followed the pantheistic philosophies of Baruch Spinoza, who considered God and nature to be synonymous. Panentheism has a slightly different meaning, accepting that God exists in creation yet also transcends it. Bear in mind that ancient people and those today may interpret things uniquely and not everyone fits into a set definition.

Pagan Parade

Since all societies were, by definition, pagan before Christianity, and those ancient cultures were agriculturalists, I see many pagan activities still being practiced among the agricultural people of La Plata County today, especially during holidays and activities such as the annual Independence Day parade.

"70% containment — and rain!" read the headlines on July 4, 2002, of the *Durango Herald*. After weeks of fire and devastation, firefighters had subdued the blaze in some areas; the communications towers remained standing; and the showers that settled the dust on July 3 brought a feeling of relief and hope to residents. Even though the fire danger prohibited the use of fireworks, moods improved and spirits lifted for the Independence Day celebration.

Jean-Luc and I visited the small town of Bayfield for the annual parade. We chose to attend Bayfield's festivities not for fancy floats and large marching bands, but exactly for the opposite reason. This parade featured house pets, horses, kids on bicycles, and senior citizens carrying flags.

A long line of families in festive clothing waited outside the Lions Club hall for the pancake breakfast. Inside, the odor of sausage and syrup sweetened the air. Earl, one of the original founders of the Bayfield Lions Club, poured coffee at the far end of the food tables. He had been one of the people who organized the first Bayfield Fourth of July parade.

Jean-Luc laughed. "Look at him, kissing every woman in line."

"You must be the kissing committee," I said to Earl as he kissed me.

We ate our pancakes and sipped orange juice at a long banquet table squeezed in between ranching families, and then we walked over to Mill Street for the parade. On the way we met Larry Sawyer, a friend and client who fells trees and skids the logs using draft horses. "Hey, Sawyer, I heard you got caught in the West Valley fire," said Jean-Luc.

Larry's large smile covered half his face. "Yeah, that was scary. I was up at the Falls Creek subdivision clearing timber around a house. The fire was on the other side of the valley, so I figured we were safe. Then I saw a helicopter come over the top of me and drop water right behind me. I looked up, saw flames, and knew I needed to get out of there. The only problem was I had felled trees all around myself and the truck — I was blocked in. I grabbed the saw and madly started cutting and rolling pieces out of the way. The horses were great; they were completely calm. They loaded right up, and I drove off. Then I took a curve too tight and drove into a ditch. I had to unload the horses, pull the trailer out, and reload the horses. They were wonderful; they seemed to know that something was up, and they behaved perfectly. It was close; the house I was working around burned to the ground."

Festivities started with the "pet parade." Dogs with American flag bandanas around their necks pulled wagons carrying tots waving flags. A skinny man dressed like Batman rode a motor scooter and handed out toothpaste. The main parade began, as it traditionally did, with a crop duster plane flying low over Mill Street. Everyone cheered, and then came the veterans carrying flags and singing "The Star-Spangled Banner."

There were kids on ponies and women dressed in saloon attire on Arabian horses. Old ranchers sat on wagons pulled by mules, and cowboys drove chuck wagons drawn by draft horses. Then came the Shriners driving miniature cars in circles. They wore red hats and sashes with emblems containing Egyptian symbols, such as the sphinx and pyramid. Next in line came the four-wheeler club, with jeeps that crawled on top of each other, followed by the fire trucks.

Usually, fire trucks were appreciated most by little boys, but not this year. This year people rose from their lawn chairs and street curb seats to give heartfelt ovations for every fire truck and firefighter present, and there were a lot of them, rows and rows of tired young men and women. Even as they were stopped because of the slow parade, people continued to stand, clap, cheer, and cry.

Behind the firefighters came the wives of the Cattlemen's Association, the "Cow Belles," driving late-model pickup trucks and waving their muscular arms at the crowds. One such truck displayed a sign that said, "Hanging Tough in Tough Times." I pointed at two yellow, moldy bales of hay in the pickup bed, which the women sat on. "Look at the hay in the back of the truck."

"Yep, they'll be selling that later for twelve dollars a bale," said Sawyer. Hay was scarce, and prices had gone sky high.

Almost every float, car, or truck in the parade had a dog in it. There were

Dalmatians on the fire trucks, Boston terriers in the semi-truck cabs, and blue heelers on the cattlemen's pickups and flat beds. The cars had poodles, the kids walked with Shelties, and the officers marched with German shepherds. Each dog was elaborately decorated with red, white, and blue accessories around their necks. Their mouths were open wide with grins of joy as they showed themselves off, thrilled to be part of the action, and companions to their masters.

According to Ken Dowden in *European Paganism*, early Christians were told in sermons to not participate in a long list of pagan behavior: don't hang things around the necks of animals or people, even Christian objects; don't pay attention to birdsong when on a journey; don't celebrate by dancing and drinking; don't take holidays on Thursdays (July 4 was on Thursday that year); don't dress up as animals like shamans and pagans; and don't make Egyptian days out of favorable days.[54] All those decorations around the necks of the dogs and horses, the bat costume, and the Shriners' emblems with the sphinx reminded me of ancient paganism.

The Cow Goddess

Although I rarely treat cattle, I enjoy observing them. One day, as I drove down Elmore Road, a black calf jumped out in front of the truck. I stopped and pulled over, hoping to move her away from traffic, and when I opened the truck door, she ran into the ditch and crashed into the fence. It was hog fence, and there was no way through it. Concerned for her safety, and that of passing motorists, I herded her toward the gate, wanting to drive her through it; however, below the gate was a cattle guard.

Cattle guards are made of pipes laid side by side with open spaces between them over a hole in the ground. They are built to allow automobiles to drive over them while keeping livestock from walking out an open gate. Indeed, if horses and cattle attempt to cross and their legs slip between the pipes, it can cause severe damage to their limbs.

The calf stood on the road that led through the gate and over the cattle guard, and any wrong move on my part could be fatal for her. She wanted to go into the pasture to join the herd. Over the center of the cattle guard was a piece of metal mesh about a foot wide, just big enough for a person to walk on. I stood on it, opened the gate wide, then moved away from the heifer. She looked down, found the mesh, and carefully stepped onto it and over the guard, safely into the pasture. I was relieved. I had never given such a beast the mental credit she was apparently due. Something smarter than me was taking care of that calf.

I have also long puzzled over what appear to be "calf nurseries." In the spring, the cows go out to graze, and all the calves lie down together with one or two cows. I started asking cowboys: "Which cow babysits the calves?" Everyone knew what I was asking about.

The first man I questioned — Tom McGee — taught gentle herding of cattle. "It isn't the top cow, she's out eating," he said. "And it isn't a cow on the bottom of the pecking order because she's too meek to defend the calves from predators. And it's not a heifer either; she's too inexperienced. It must be a cow in the middle of the pecking order."

Cattle are among a number of animals that share the responsibility of caring for the young and have "nurseries" or "kindergarten" groups of calves. Burt Smith mentions in his book on low-stress handling of cattle, *Moving 'Em*, that occasionally, the bull does the babysitting of the kindergarten.

I attended to a horse belonging to cattlemen Joe and Valerie Duke, and when I asked them the same question, they started laughing. Joe said, "They'll leave their calves with anyone that will stay with them, including Buford."

"Who's Buford?" I asked.

"Buford is the mechanical roping steer. We came home one day to find all the cows out eating, and all the calves were lying around Buford."

Another cattleman insisted that it depended on the qualities of motherhood. "That cow there always loses her calves; I need to sell her. The good mothers stay behind, and the bad mothers go off and eat."

Cattle have taught me one thing, and that is, no matter how far away the cows are, if I walk into an area with calves, especially if I have a dog with me, the cows come bellowing to the aid of their calves, keeping their eyes glued on me or the dog. The cows don't just go off and completely forget about their young.

I heard another interesting answer. I was working as the veterinarian on duty at a Skijoring competition. In this sport, a horse and rider pull a skier who skies through gates, goes over jumps made of hay bales, and snags rings on their arm, while the horse runs flat out, straight down the center of the course. The skiers crash a lot, but the horses rarely get hurt, so as a vet at a Skijoring, I spent a good deal of time chatting with spectators. Lucky Simons approached me, real close.

"How are you, little lady?"

He smelled like his paper cup, which contained whiskey. He was nearing retirement age, but he was still handsome and soft-spoken. He said he had been moving a few yearlings lately, so I asked him, "You know those calf nurseries?"

"Yeah, isn't that something?"

"Which cow stays behind and babysits?"

"Well, haven't you ever seen them draw straws? They probably ask themselves, 'How do those people know how to go into that bar?' They talk among themselves — instincts. It's not a heifer... maybe they don't know enough yet. It can sometimes be an older cow." He had never seen a bull babysit or a mechanical roping steer. Lucky agreed with the importance of low-stress handling techniques, although he had never read a book on it. "We don't allow any whips or yelling; you can do your yelling in the bar. Gentle handling is easier on the pens, the horses, you, and everything else."

"Yes, like rate of weight gain, milk production, and fertility," I added.

"That's right." Although this cattleman was a Christian, he gave the Goddess her due. He summed things up by saying, "There are a lot of amazing things you can see if you watch animals. God and Mother Nature got it all figured out."

CHAPTER 5

Hinduism: Escape from Samsara

Reality itself may be only a symbol for the workings of God's mind.
And in that case, the primitive belief,
found throughout the ancient and pagan world,
that God exists in every blade of grass, in every creature,
and even in the earth and sky, may contain the highest truth.[1]
— DEEPAK CHOPRA

All religions evolved from paganism and shamanism, and they still contain many of their symbols and beliefs. Even today, Hindus revere the Divine Mother principle along with the Father God, just as shamans and pagans did. Historically, agricultural communities living in the Indus River Valley worshipped the Great Mother Goddess until around 1500 BCE, when they were raided and conquered by patriarchal Aryan warriors who moved in from central Asia. These pastoral nomads charged in on horseback and light chariots, devastating the "witches," as they called them.[2] The invaders herded cattle and sang hymns to their sky god, as well as to the forces of nature — the sun, wind, lightning, and fire. They conquered the local Dravidian civilization, assimilating the people and their religions, creating a blend of pagan and patriarchal religious beliefs. A synthesis of philosophies resulted, creating a collection of spiritual ideas we call Hinduism.

A Dream in the Mind of God

The word *Hindu* refers not to a religion but rather to a geographic location. It is a Sanskrit name for *Indus*, meaning "ocean" or "river." Hinduism has multitudinous sects, and it resists a single definition. Hindu ethics permits a wide range of religious beliefs; hence generalizations are difficult to make. However, three predominant concepts must be discussed about this fascinating religion — monotheism, reincarnation, and karma.

Hinduism is strictly monotheistic, dominated by the concept of monism — the oneness of all things. One God exists in a reality unknown to us, and *our* entire reality is like a dream in the mind of God. This reality is an illusion created with our senses; the truth is oneness with God.

As the nineteenth-century Indian mystic and yogi Ramakrishna explained, "When a man is on the plains, he sees the lowly grass and the mighty pine tree and says, 'How big is the tree and how small is the grass!' But when he ascends the mountain and looks from its peak on the plain below, the mighty pine and the lowly grass blend into one indistinguishable mass of green verdure. So in the sight of the worldly there are differences of rank and position — one is king, another is a cobbler, one a father, another a son, and so on — but when the divine sight is opened, all appear as equal and one, and there remains no distinction of good and bad, high and low."[3] This concept of making no distinction is called "equanimity," another prominent teaching of Hinduism. It is our true nature, a nonattached acceptance, pure awareness, when one transcends duality and finds union with everything.

For Hindus, the whole universe is a projection of God's nature. Where God came from (or what the Supreme God is) is indefinable. Although there are 330 million Hindu gods, they are all part of the one that is divine.[4] The "gods" are anthropomorphized principles of the one that is unknowable, without sex — beyond pairs of opposites. "Truth is One; they call him by different names," states the Vedas.[5]

The Vedic texts are believed to be one of the world's oldest written scriptures, originating from the hymns sung by the Aryan nomads, developed circa 4000 BCE. Their chants told of a beginning with neither being nor nonbeing, neither death nor no death; there was only the undifferentiated principle called "One."

The ancient sages noticed that everything in existence was impermanent and ever changing, and some invisible source of life prevailed behind the temporary physical forms. In the Vedas, that source is called Brahman — the pure, invisible, unchanging spirit.

The physical world (*maya*) is in Brahman, and in everything there is a part of Brahman, called Atman, or Universal Self. The Atman is the eternal and unchanging essence of everything that is a part of God. The Vedas recant a universal theme: "The Truth is within us."[6]

The Hindu sees God in everything; even the lowliest creatures are sacred. Each contains the Atman, the soul that is pure spirit. In India, the soul has another part besides the Atman. The jiva corresponds to an individual's soul making its journey through many lifetimes until it reaches full realization of God. Atman cannot change in any way. It never reaches God because it never left in the first place. The jiva, on the other hand, may require many reincarnations to find God.

The philosophy of reincarnation developed from observations of nature. Ice melts to water and evaporates away as steam only to condense into water and freeze again. One sees that nothing in existence is ever absolutely destroyed; things merely change form. Thus, the life force that is present in the material world must transmigrate or, in the case of creatures, reincarnate.

The Vedic literature refers to transmigration of souls as *samsara*, a doctrine of unknown origin, although many tribal beliefs worldwide have long embraced the concept, as did a number of early civilizations, including the druids, the Greeks (Plato and Pythagoras), and African tribes.[7] The aim of Hindu spiritual practice is escape from samsara, the illusion of the material world and the cycle of death and rebirth, to find unity with God.

Hinduism was formed as the idea of transmigration of the self was synthesized with the belief that God is in every part of reality and the unknown sky above. All of the previously discussed religious ideas — shamanism and paganism — melded together with patriarchal thought and the concept of rebirth. Since tribal beliefs held that animals have souls, and since God is in everything, it followed that animals and humans alike were subject to the laws of reincarnation. In order to understand these laws, we must first understand the concept of karma.

Karma is the law of cause and effect by which beings determine their own fate. You reap what you sow; these are the laws of morality. Together with the principle of reincarnation, this means that you may reap what you have sown in a former lifetime. Each being goes through countless lifetimes, in various forms, in order to transcend, eventually becoming one with God. So a human may have been a rock, a tree, an aardvark, a dolphin, a king, a demon, and so on. The important task remains the evolution of the individual jiva, until one ultimately attains liberation from this physical illusion.

In my daily life as a veterinarian, I see many animals suffering. Often, their lives are out of my control, and the sight of them saddens me. I once often visited a ranch that had a dingy rental trailer on the property, where a very small cage held a husky dog that lived among piles of its own dried feces. The stench was disgusting, and the dog appeared mentally vacant. Year after year I passed within several feet of this cage, and the dog never made eye contact with me; he seemed fragile and never acknowledged me. He lay under a covered area never attempting to rise. I believe that if I had opened that cage, the dog would not have left. It had food, water, and shelter; the law considered that adequate care, and I saw no way to help the dog other than to pray. My only consolation was thinking that perhaps in a previous life that dog was a dog beater whose karma created his return as a dog destined to sit and contemplate the life of a dog.

Sathya Sai Baba (1926–2011) was an Indian guru and spiritual leader who was said to perform miracles. A story I heard about him opened my mind to the purpose of karma. One day Sai Baba was walking through a crowd with a number of disciples, healing people along the way. A disciple questioned him about why he sometimes healed old people and passed over young children. Sai Baba answered with a question: "Can you see their karma?" The disciple understood that he could not but that Sai Baba could.

Perhaps there was more going on with the jiva of the dog than I realized. Maybe the dog was exactly where it needed to be. I could only pray that he received the healing he wanted, ask that his karma be forgiven, and that his next life be much better. Also, I had to reflect upon the parts of myself that resembled that dog, unable to escape its suffering.

Originally, the idea of reincarnation gave people hope of ascending from lowly life to oneness with the divine, while later it became a threat of punishment for bad behavior. For example, in one group of morality tales, a foolish man is reborn as a monkey, a cunning one as a jackal, a greedy one as a crow. Conversely, an animal may rise to human status, in stages or all at once if it has done the right deeds. A life well lived brings rewards in the next incarnation.

A person who believes in reincarnation may take this literally, assuming that an animal may indeed be someone they once knew. Another interpretation is that the human is reborn to an irrational existence comparable to animal life, and not actually attached to the body of an animal. For example, a cheating person is a rat or a snake. Or one could accept that both may occur, since a wide range of interpretation is acceptable. Hinduism has a great capacity for absorbing ideas.

Although some authors describe one animal as more spiritually evolved than another — for example, a giraffe being more spiritually evolved than a grasshopper — the Hindu literature does not specifically indicate a line of spiritual superiority among the animals. Humans believe that animals are less evolved spiritually than humans, but there may be exceptions.

The holy trinity of the supreme God in Hinduism is Brahma the creator, Vishnu the savior, and Shiva the destroyer. Krishna is the ninth avatar, or incarnation, of Lord Vishnu. He says in the Bhagavad Gita, "I am the conscience in the heart of all creatures." Krishna teaches that all species of life in this material nature live in one of three modes of existence: goodness, passion, and ignorance. He says, "One who dies in the mode of passion takes birth among those engaged in fruitive activities [capitalism, for example], and one who dies in the mode of ignorance takes birth in the animal kingdom."[8] A human could also take birth in an irrational human form (acting like a pig in the mode of ignorance).

Although human existence is generally regarded as higher than that of a dog, for example, some speak enviously about the position of the dogs that live at the base of Mount Govardhana and daily drink the milk *prasad* from this natural form of Krishna. The assumption is that these dogs have achieved a high spiritual state.

Krishna calls himself "the holy fig tree" as well as a "lordly elephant" and a horse "who rose out of the ocean."[9] In other words, he is in everything and shines as the holiest of each particular species or material, such as the "milk prasad," which indicates that some animals may be more spiritually enlightened than some humans.

Animals are subject to slightly different laws, according to Swami Prabhupada: "The Vedic literature is meant for human beings, not for animals. An animal can kill without sin because he is bound by the modes of nature. But if man kills, he is responsible. He has a choice in his actions."[10]

However, Krishna does not indicate that all killing is wrong. In the Bhagavad Gita, Krishna tells Arjuna, a soldier who accepts Lord Krishna as the "Supreme Personality of Godhead," to fight his kin as a matter of duty. Krishna says, "Therefore, O Arjuna, surrendering all your works unto Me, with mind intent on Me, and without desire for gain and free from egoism and lethargy — fight."[11] And since the soul is invincible, you should not grieve for the body. Arjuna had to be fully "Krishna Conscious" in order to be free of karma, and he had to fight because of his position in life. This would imply that a military soldier with orders to kill, if his or her heart were in the right place, in a state of equanimity, would not have bad karma attached to his or her actions.

Bad karma may degrade a human to animal form, according to Prabhupada. Someone who eats meat and other "abominable foods" comes back as a hog in the next life; a man lusting after a woman comes back (via a "stinking womb") in a dog's body. An animal must then elevate itself through the evolutionary process to reach human form again. Regardless, Krishna is inside (the heart) of all beings, directing the wanderings of all living entities, and he is eternal. "Therefore, you need not grieve for any creature," writes Prabhupada.[12] There is no death for any of us. Furthermore, any being can rise above its current mode of existence, which is a matter of karma.

Both good and bad karma bind us and must be overcome in order for one to be liberated from this earthly existence and transcend to the Ultimate Reality. Karma, unlike sin, has no moral blame attached to it. Karma is any act that leaves an impression; it is action with intention. Karma can be right or wrong and still leave an imprint. One transcends karma by working without attachment to the fruits of labor in full devotional service and by gaining spiritual knowledge. Krishna advises "aversion to fault finding" and "compassion toward every living entity."[13]

The law of karma also includes the blessings of grace. Divine authority can forgive karma. One way to achieve grace is by reciting mantras — sacred Sanskrit words. Mantras work energetically on the body's energy system to relieve past karma.

When I first read about this, it sounded like a good idea to me, and I started chanting mantras in the truck as I drove around La Plata County, listening to the tape series on Hindu mantras called *Mantra: Sacred Words of Power* by Thomas Ashley-Farrand. When frightened, riding upon my mule on steep cliff sides, I sang a Durga mantra for protection, "Om Dum Durgayei Namaha." While attending the ill, I sang for healing, and as I irrigated Earl's pasture, I sang Rama mantras to realize the divine Self: "Om Sri Rama Jaya Rama Jaya Jaya Rama." I noticed immediately that I felt happier. In place of worried thoughts, I chanted beautiful words. Instead of judging, I enjoyed vibrational tones. Mantras provided a method of self-improvement while avoiding detrimental mental activity.

Animals use a lot of sound vibrations — purrs, hisses, roars, squawks, squeals, and melodic songs. Perhaps these noises work like mantras to alter their energetic bodies. The world began with sound; from that powerful vibrational energy came light and everything else. Animal vocalization could be used as an energetic device for creating safety or healing.

One Buddhist tale says, "The peacock escapes danger by reciting every day a hymn to the sun and the praises of past Buddhas."[14] In addition, evidence suggests that the purr of a cat may have healing qualities. Cats purr at frequencies of 20 to 50 hertz, which are the same frequencies that are proven to be the best for fracture healing and bone growth. The purr may have an effect similar to ultrasound on the body. *The New Zealand Veterinary Journal* mentioned a case of a cat with severe dyspnea who could breathe more easily when he purred. Other medical literature mentions the positive effects of chest-wall vibration in cases of obstructive pulmonary diseases, as well as the effects of stimulus frequency on tendons and nerves and on bone healing, which is in part how electro-acupuncture works.[15]

Common sayings in veterinary circles are "If you put two cat bones in the same room, they heal," and "cats have nine lives," since cats heal well with little human intervention. A cat's purr resembles the sound of the mantra "Rama, Rama." Perhaps the vibrations from the purr at 20 hertz work on the energy system of the body to bring about healing, just as mantras are supposed to do.

We know an energy system exists in the body, and modern medicine measures it with electrocardiograms, electromyograms, and electroencephalograms — the Chinese call it "qi." Hindus say the body's energy system includes the following: the aura, which is the energy around the body (witnessed as halos); the chakras, which are the vibrational centers situated along the spine; and the energetic channels that flow through the body.

In humans, seven chakras of different colors exist along the midline: the first (red) at the base of the spine, the second (orange) below the navel, the third (yellow) in the solar plexus, the fourth (green) at the heart center, the fifth (blue) in the throat, the sixth (purple) at the brow, and the seventh (white) sitting at the top of the skull — the crown chakra. Animals also have chakras, I am told, however, they are more rudimentary or perhaps they have fewer; it depends on whom you ask. According to my clairvoyant friend Dana, beings in animal form learn from each incarnation how to manage the energy of a particular body, eventually evolving into more complicated bodies with more challenging energy systems. Animals also have energy channels, as one veterinarian demonstrated at a conference I attended in China in October 2000. She placed an acupuncture needle into an acupoint and touched it with a vibrating tuning fork to measure the acoustic resonance to mark the channel.

To further explore some concepts of Hinduism, consider the following story from La Plata County.

Rebirth

The first time I saw Lonna Robertson, her water-blue eyes looked into me as if she already knew me. "Are you Karlene, the veterinarian?" she asked. I admitted to it, and she told me she might want me to put her old dog "to sleep" in the near future.

The day came, and I met Michael, a geriatric husky with a powerful presence. He made his last climb up a step stool into the back of Lonna's truck, his favorite place, ready to pass with dignity and grace. I injected the euthanasia solution into his right cephalic vein as Lonna held him, and she felt his body relax as he left. "This is easier for me now that I know where he's going," said Lonna. "I know because I died once; I know how wonderful it is."

"Really, can you tell me...while you were dead, did you see any animals?"

"I did. I saw horses," she said, and she told me about her near-death experience.

It was Saturday, February 12, 2000. Marshall and Sally Goldberg were in New York City at the Westminster Dog Show with their standard poodle, and they had hired Lonna to watch their home, their horses — Mindy and King — and the Goldberg's other dog, a greyhound named Beth. About 5:30 PM, a neighbor called to tell Lonna that the Goldbergs' horses were out on the road. Lonna explained that she had just fed the horses, and she knew they were in their paddock. "They must be someone else's horses," said Lonna. "What color are they?" The woman said that they were white. Since King was white, Lonna decided to investigate.

As she walked down the curved driveway to the barn, she could see two white horses running down the county road toward her. They were slipping and sliding on the pavement in six inches of fresh snow. Since horses are attracted to other horses, the two white strays came onto the Goldberg's property to meet Mindy and King at their paddock, expressing their greetings and opinions with much grunting and squealing.

Lonna opened a gate to the adjoining pasture, trying to lure the loose horses into a place where they would be of less danger to themselves and passing motorists. She made sounds like whinnies and nickers in an attempt to call them.

In the dusk, the colors of the world mellowed, diminishing visibility; flakes of snow as large as white aster flowers fell onto her long blond hair and onto the two horses as they moved toward the open gate. Lonna was careful not to get behind them, but as she closed the gate, pushing through the snow, the white mare kicked at her and connected with the inside of Lonna's right thigh.

The intensity of the pain was enough that Lonna thought her leg was broken. She tried to close the gate and found that she was unable to move her leg. Her foot felt warm and wet in the rubber boot, so she looked down. Through a hole in her jeans and long johns spouted a fountain of blood spewing from the femoral artery, pulsing with the rhythm of her heart. She knew she only had seven to fifteen minutes of life and attempted to calm down and cool herself by dropping to the ground and packing her leg with snow. She felt sedated, the world slowed, and her vision faded. Now, without sight, her only hope was to crawl to the county road where someone might find her.

The previous October, her astrologer had warned her to be careful because February would begin a dangerous time when Chiron squared the sun in Lonna's chart. Chiron, infamous for accidents, is a death archetype. Often symbolized as a white centaur, Chiron is like a poke with a sharp stick in an open wound. It foretold that she might have to face her fears and possibly choose between life and death, either a shamanic death or perhaps an actual death. In either case, Chiron often initiates a wounding and healing cycle, a cathartic cleansing, pushing one to grow.

She remembered all this as she was dying in the snow on her hands and knees, dragging herself to a quiet country road, hoping for a miracle to save her. The sense of touch, her only remaining perception, guided her along the gravel driveway as she searched for the pavement and tried to avoid the deep ditches that bordered both sides of the driveway.

Because of the pain, she stopped several times, leaving puddles of blood behind. Each time she stopped, the farther away she drifted from life, and the more peaceful she became. Then the fears and pain waned, and she surrendered, collapsing into the sweetest emotion she had ever known. It carried her to another place, enveloped by white light and bathed in serenity. There, she knew that everything was as it should be. In a rapt bliss, she saw coming out of the light a herd of white horses running wild and free. They roared over her, but she had no fear; rather, they wanted her to follow them, and she was swept along with them, sailing amid their manes.

Meanwhile, George Bruekner and his wife, Rose, were on their way to feed a friend's chickens and gather eggs. They drove slowly because, according to them, there was a foot of snow covering the road. As they passed the Goldberg's, they saw Lonna, still in the driveway, about fifteen feet off the road, on her hands and knees with one arm outstretched before her, waving to them.

When George scooped her into his arms, Lonna felt as though she was slammed back into her body, and her senses were revived. Pain and fear returned, and she could see the word *Bronco* on the vehicle. George placed her in the front passenger seat, and Rose held her from behind. Back in the world of the senses, the stench of cigarette smoke nauseated her, and she started to worry about her dog, Michael, and Beth, alone at the Goldberg's.

Blood splattered onto the emergency room floor as they cut away the rubber boot. Lonna's heart had stopped. Yet she saw herself and all the chaos around her in what seemed like a large mirror. Hysterical nurses tried to get a blood sample for blood typing from an arm vein, until Dr. Denver entered and said, "Take the blood from the wound! Take the blood from the wound!" Even after the sample was collected, Lonna saw one nervous nurse accidentally throw the vial into the trash, but she could not speak; she was gone, flatlined.

CPR brought her back for a short time. Lonna heard the nurses asking, "Who shot you?" "Why are you so skinny?" But she tired of the chaos, and having lost half of her blood volume, she let go again, surrendering into the peace of the other world. The line on the ECG went flat a second time. As though she were half in one world and half in the other, she continued to hear comments in the emergency room and see what was going on but remained separated from it. People said she would not survive, and they continued to remark about how tiny she was and that someone had shot her.

Then Lonna felt the presence of dead family members telling her that all was well, and she sensed a support group of other beings. She bathed in a brightness accompanied by the sense of serene perfection. Musical chants and colors appeared that she had never known before, and she was given a choice: Do you want to return to life or not?

Her concern for the dogs, and the dream of a radical snowboard trip she had planned in Montana, enticed her. The Goldberg's dog, Beth, was in her last days at this time, and Lonna had been massaging her and administering medications. Lonna thought about Beth and how she loved her, and how much Marshall Goldberg loved his pet. She thought about how much she loved Michael, who would be wondering where Mom was.

In Lonna's mind, regardless of the blood transfusion, the heart massage, and the electroshock used by the emergency medical team, ultimately it was Lonna's decision, and she chose to stay in the physical world. In that moment, she channeled the energy to return again with a slam into her body.

In preparation for surgery to repair the section of damaged artery, the doctors and nurses continued to discuss Lonna's condition, not realizing Lonna could hear them. They mentioned the possibility of amputating the leg. That jolted Lonna, who frightened a nurse as she awoke, grabbing her to say, "Cutting off my leg is not an option. You do not have my permission to cut off my leg."

Dr. Rosenbaum prepared to anesthetize Lonna, but she insisted that someone check on the dogs and horses first. Once Dr. Rosenbaum convinced her that her boyfriend was on his way to take care of the animals, she agreed to anesthesia. Once she was ready, she asked him, "Do you think I'm going to make it?" He said, "You're strong, you made it this far. So, I'll see you when you wake up."

Dr. Rosenbaum acted like an anchor for Lonna in the turbulence of the hospital scene. He held her feet or hands or touched her head. His tender, calm demeanor provided a pleasant contrast from the emergency room. He grounded her and provided a comfortable place to return to. She stayed focused on his loving, caring energy as a reminder of how wonderful life is, and she confirmed her decision to stay here, since she could at any time have chosen to go back to death. "His energy was as sweet as the energy on the other side," she emphasized.

In the weeks to follow, Lonna became obsessed with the idea of owning a sword. That idea faded, and with it the sensations she felt on the other side. She ached to remember the surrender and peace, and that all is well. Instead, she remembered what life presents: struggle, fear, and pain. But now she felt grateful for the wonders of living.

Lonna cried as she told me that putting Michael to sleep was the hardest thing she had ever done. Yet the experience reminded her of her own death, and she realized how insignificant all the worries of the world are. She explained that by participating in "Michael's rebirth," she again tasted the sweet memory of surrender. She suggested that when other people face death with their pets, they gain perspective on what really matters — love — and trust that all is well. Lonna's love of animals brought her to death and back to life again. Love was the feeling — the vibration she held on to.

Lonna told me, "Regret, what if, should have, blame, sadness, pain, shame, guilt, hate — none of those are there on the other side. You really have to be strong to get through life and be positive about it. Life is hard, but we can help each other. Life can make you or break you. It's much easier to die. The easiest thing I've ever done was dying. Perhaps it is our challenge in this life to get past the fear and learn to walk this earth with an open heart."

Lonna's words are echoed by many other people who have had similar experiences. According to physician Melvin Morse, near-death experiences are not caused by a lack of oxygen to the brain, drugs, or psychological stresses evoked by the fear of dying. Almost twenty years of scientific research have documented that these experiences are a natural and normal process. We have even documented an area of the brain that allows us to have the experience. Near-death experiences are absolutely real and not hallucinations of the mind. They are as real as any other human capability. Morse says, "I am struck again and again that those who have entered into God's light at the end of life return with a simple and beautiful message. 'Love is supreme.... Love must govern.... We create our own surroundings by the thoughts we think. We are to love one another.'"[16]

The near-death experience speaks like a dream or vision and is common across cultures. Individual experiences are interpreted with images from the person's symbolic dictionary. One person may see Christ; another might see the White Buffalo Woman. Lonna saw white horses.

Horses symbolize many things, most commonly freedom and power. White horses often symbolize peace. They are mentioned in a number of religions, including Hinduism and Christianity, and both mention them in the same circumstance, which requires some background to explain.

In Hinduism, the supreme divinity is often described as a trinity, as it is in Christianity: Creator/Brahma, Savior/Vishnu, and Spirit/Shiva. The savior aspect of that trinity is Vishnu, "The Preserver." Hindus believe that Vishnu has appeared in physical form on the planet to help humanity nine times. These nine incarnations are called avatars — spiritual beings who come to earth to help people.

The nine avatars of Vishnu are: 1) a fish that saved humankind during the great flood; 2) a tortoise that grounded the floating landmass of India to the continent; 3) a boar that killed a demon and saved us from another geological disaster; 4) a man-lion that also defeated a demon; 5) a dwarf who overcame an evil king; 6) Parasurama, the ax-wielding Rama, who came to chop off the heads of evil kings; 7) Rama, the ideal hero; 8) Krishna, the supreme personality of the godhead; and 9) the Buddha, the vehicle of absolute Truth.

A tenth avatar — Kalki, "the coming one" — is yet to come. From the *Vishnu Purana:* "Kalki will ride a white horse. His one arm will be held high and there will be a fiery sword, like a comet in his hand. Kalki will bring about the final destruction of the wicked world, and then a new creation will begin. His greatness and might will unobstructively prevail. He will restore purity and virtue to the world."[17] He is the savior coming to harvest the souls waiting for liberation. Another interpretation is that Kalki will be born as a white horse, the animal and the god being one in bodily appearance.

In Christianity, a similar legend holds that the savior, who is Christ, will ride a white horse in heaven, with a sword, and create a new heaven and a new earth, as it is described in Revelations 19:11–16; 21:1–2.

The white horse is a vehicle, or a symbol, of the savior who purifies life, vanquishing evil and restoring purity. Even the vehicle that Lonna rode to the hospital was a "Bronco." The "savior" is recognized as Vishnu, Christ, or a white horse; it is an archetype, an interpretation from a person's symbolic dictionary. Some believe that Christ is Kalki; in any name, they are the same energy — love, purity, the one who triumphs over evil, the "God" who saves us riding a white horse and carrying a sword.

According to Thomas Ashley-Farrand, a Vedic priest: "Any teacher who comes and teaches us something about spirituality or how to solve worldly problems in a spiritual way contains the energy of Vishnu. That principle comes in all the spiritual teachers, in all the religions and all the paths. Because we come from so many cultures and states of mind, the truth has to come in many different forms so that we can grasp it."[18]

In her book *Embraced by the Light*, Betty J. Eadie explains that she saw Christ during her near-death experience. She describes what the son of God told her: "Each of us, I was told, is at a different level of spiritual development and understanding. Each person is therefore prepared for a different level of spiritual knowledge. All religions upon the earth are necessary because there are people that need what they teach."[19]

Her message from Jesus Christ is the exact practice accepted in Hinduism — an acceptance of a wide range of religious philosophies. Hinduism even allows the practice of worshipping images of the gods. From Ramakrishna we learn that, "though the educated Hindu rejects such practices for himself, he nevertheless looks on them with a kindly tolerance because to him every form of worship, no matter how crude it may be, is a stepping stone to a higher form."[20] Furthermore, it is not for us to judge.

Hindus have long understood and described the process of leaving the body via the seventh chakra at the top of the head. For centuries, mystics have witnessed it without a traumatic near-death experience. Thomas Ashley-Farrand describes what it is like to leave the body in his discussion of the seventh chakra, which is the connection to the All: One pours the cup of one's consciousness into the ocean at the seventh. At this point, your consciousness enters the ocean of consciousness. And for a time, you may be at oneness, not even aware of your separateness. And when you come out, you are aware — *oh, yes, I have a body; I exist as a separate individual*. But at that point you have poured your cup into the ocean of consciousness, and then when you come out, you dip it back in again. And they say that those who experience that are never the same when they come back to normal waking consciousness.[21]

Lonna was not the same. For her it was a cathartic purification, a new beginning and rebirth. I learned from Lonna that death is not painful, that the rider of the white horse holds us, as many sacred texts state. When an animal dies, it does not suffer terribly because the spirit has left the body and is held in bliss.

The important thing to remember is that love holds a life to this reality better than medicine, better than surgery. In the case of a dying pet, it is sometimes best to love the animal as it passes rather than hysterically rush about in lifesaving efforts.

Animal Vehicles and Sacred Animals

The Hindu gods are often depicted riding animals, which are their vehicles or *vahanas*, each representing a divine attribute. Myths about these gods and their vehicles teach the moral principles they represent. Male gods have female counterparts or "consorts." Hindus believe there can be no male without the female; the male part provides the idea, while the female part provides the power.

The female consort of Brahma is Saraswati, goddess of learning. Saraswati rides a swan. The swan represents the two aspects of nature for all living beings. Although it swims on the surface of the water, the swan is not limited to water; it can fly into the air, at ease in both places, ever free and unconcerned with the events of individual life.

The consort of Vishnu is Lakshmi, goddess of wealth, a beautiful woman pictured among two elephants with raised trunks. The elephant represents the most valued animal of India, serving as a military vehicle, a laborer of great strength who is able to negotiate difficult terrain and yet is gentle and docile.

Durga, the goddess of protection, the invincible power of nature, rides a tiger, reminiscent of the Mother Goddess art of the Sumerians and Assyrians. The tiger is one of Durga's many weapons.

The most popular god may be Ganesha, the elephant-headed god, the pot-bellied, jolly son of Shiva and Parvati, the god of overcoming obstacles who brings about unity. Ganesha's vehicle is the rat — the symbol of the ego. Like everything in this world of duality, the ego presents both positive and negative components. On the positive side, it provides the power of creativity. Without ego, there is no invention.

The epic poem the *Ramayana* includes a story about the monkey god, Hanuman, sometimes referred to as an avatar of the god Shiva, whose name implies "one whose pride was destroyed." Hanuman served as a faithful servant to Rama. Rama's wife, Sita, had been abducted by an evil demon. Hanuman, a fast, devoted monkey, ran through the trees to find where the demon had taken her, then ran back to tell Rama. Thus, Hanuman symbolizes energy and strength. The *Ramayana* also mentions that "monkey-heroes were sons of gods begotten by simian females for the express purpose of helping Rama."[22] Hindus revere monkeys because of the belief that a monkey or ape may be a degraded human.

Almost every animal is sacred to Hindus, in particular the cow, the national animal of India. The cow is considered to be the axis on which the economy of agriculture revolves. Cows are important economically as a source for milk and ghee (clarified butter), as well as for pulling carts and for fuel (cow dung). Poverty in India is blamed on the lack of protection of the cow. Not only is the cow sacred but also her five products — milk, curd, ghee, dung, and urine.

Lord Krishna is depicted with the cow, having been born as a cowherd; he is the guardian of cows. As we learn from pagan belief systems, the cow is the symbol of the earth or the Great Mother. A number of legends tell about the "wish-fulfilling cow," the cow that feeds the many. "There is no limit to her bounty, to her patience and her compassion."[23] A cow's milk feeds a child once he or she is weaned from the human mother's milk, making the cow the surrogate mother to all children. One would never kill their mother, surrogate or not, so orthodox Hindus do not kill cows. However, Hinduism is a conglomeration of a number of beliefs and religious practices from Muslim and Christian invaders to the Parsis and Jews who came as refugees. The ancestors of some Hindus in Bengal and Mithila sacrificed and ate beef, and some Brahmins there still do.[24]

The snake, the crow, and the cow have a similar energy on an evolutionary level to humans, according to Sadhguru.[25] All temples have a snake because the snake represents an important energetic process we must recognize. The Kundalini, or primal energy of the body, depicted as a coiled snake at the base of the spine, provides a connection to universal consciousness. When asked if animals can be enlightened, Sadhguru explained that it is possible but exceedingly rare. Spiritual teachers find enough work with efforts to enlighten humans.

Ultimately, Hindus worship every animal, from cobras to monkeys, which reflects their adoring reverence or regard for all life.

The Sacred Rat

Not long after meeting Lonna, while I was reading *Sacred Animals of Nepal and India*, I had another encounter that offers insight into animal worship.

I walked in the back door of the Animas Animal Hospital and helped myself to a piece of the ice cream birthday cake that was sitting on the treatment table. Dr. Walt Truman was suturing a wound in the adjoining surgery room.

"Breaking OSHA regulations again," I said, as I reached for a tongue depressor to use as a spoon. "Well, if we veterinarians didn't eat on the job, we'd starve."

"That's right." Walt's eyes smiled above his surgical mask.

"Whose birthday, is it?" I asked.

"Adam's," said Tanya, one of the technicians.

Adam was dressed in green surgical scrubs and was lying on the floor, on his side, head propped up by one elbow, petting the clinic cat, Larry.

"How old are you, Adam?" I asked, sitting down on a stool by the microscope to eat my cake.

"Twenty-four," he replied, as he stroked Larry's long, black back.

"How long have you worked here?"

"Three years."

"I know they're glad to have you."

"Yeah, I've come a long way since I started here. I'm much better with the hands-on stuff. I can handle the blood and bones...." His eyes squinted.

"And the vomiting and diarrhea," I added.

"Yeah...and the death. There's lots of death," said Adam.

"Every day."

Adam nodded. "Yeah, we just euthanized two dogs today on a house call."

Dr. Sonya Bird walked out of the exam room and sat down to write on a record. Her long blonde hair covered her face as she wrote. She seemed sullen.

"How do you manage to keep those long, painted nails on this job?" I asked her.

"I don't know."

"My nails are short, and I have trouble keeping them clean," I remarked.

"Well, you work on horses. That's different."

"True, I performed a rectal pregnancy exam on a mare today. She was glad I had short nails." The comment did not get the laugh I expected. "What's going on?"

Her frustration rose up and ran out her long, orange fingernail as it pointed across the room to a small oxygen cage. "Fix that rat!"

"What's wrong with it?"

"It's trying to die! Look at him; he's a rack of bones."

Sure enough, the hairless, white beast was barely recognizable as a rodent.

"He has pneumonia," she said. "Whenever I take him out of the oxygen cage, he tries to die. But the people won't let me kill him. Every time they come in, and he eats a cracker or something, they get all hopeful. It's horrible. The bill is already over five hundred dollars."

"They say the Hindus worship animals," I said, "but it looks to me like we Americans treat animals more like they're sacred. I've been reading *Sacred Animals of Nepal and India*. Hindus have a festival where they worship dogs for a day. They put garlands of flowers around their necks and feed them." Dr. Bird giggled at the irony. At least I lifted her spirits a little.

We understand so little about other religions. Although the rat is sacred to Hindus, that does not imply they would take one home for a pet. Many Hindus would wonder at this treatment of a rat, especially since millions of Hindus live in poverty.

I told Dr. Bird about Lonna's near-death experience in an effort to cheer her. The rat was in divine light and not suffering. Perhaps the owner's love and "adoring reverence" kept the rat alive, and that was okay.

Before I left, I took time to pet a cat in a cage that appeared very unhappy. He had an Elizabethan collar on his neck so that he would not lick out the urinary catheter in his penis. His urethra had been blocked with urinary calculi, and he had nearly died. Now he had an IV line in his forelimb for fluid administration. He cowered as I approached him, but he quickly leaned into my hand as I scratched his head and rubbed across his eyes. "Come on, buddy, you have to purr so you can heal quickly and get out of here." He started to purr, *Rama Rama*, which made me feel better as well.

Halloween Hare Krishna

I had been trying to understand Hinduism by practicing what they preach. I took hatha yoga classes, I meditated, and I chanted mantras constantly. And for Halloween, the following year, I dressed as a Hare Krishna (Jean-Luc dressed as a guru). A seat of pantyhose covered my head of hair except at the top where my ponytail hung out, and I tied a ribbon around the nylon waistband on my forehead to make my head appear bald. I wrapped a light orange-colored sheet around me and wore sandals. In my cloth shoulder bag, I carried incense, a tambourine, and the Bhagavad Gita. I played my part, shaking the tambourine and chanting, "Hare Krishna, hare Krishna, Krishna Krishna, hare hare. Hare Rama, hare Rama, Rama Rama, hare hare."

People laughed at my costume; one man gave me a dollar; and a wizard expressed his anger at me because he had to work for a living while I just went begging. Others questioned me: "Have you read that book?"

"Yes, three times."

"Why?"

"Because I want to understand Hinduism."

"Why?"

"Because I want to be happy, and Krishna is love and joy."

It was true, and what I was reading did help me feel better in some respects. The fact that I killed animals and ate meat, which meant that I could come back as a pig, was not comforting. At least the Bhagavad Gita told me not to lament. The love imagined as Krishna lives both in my heart and the heart of a pig.

I appreciate the way Hinduism respects intuition as the only method through which the ultimate can be known. My direct intuitive experience with death gives me insight into the unifying principle behind the drama, as Lonna understood it. I hope that when people experience the death of a pet, they intuit the sacred side of death a bit more. In a sense, we put on a Halloween costume every day, creating a separate image of who we are, when in truth we are all one.

Evil versus Equanimity

In Hinduism, God is in the evil as much as the good, with the two forces battling eternally. When humans gained dualistic thinking, we divided everything into good and evil. We see dualism in the sacred rat — two words that seem mutually exclusive — since the ego, like everything else, contains both good and bad. Evil depends on perspective; religions have a long history of defining other religions as evil and seeking to eliminate them. As long as duality exists, people will convince themselves that "good" requires "doing one's duty to kill the other."

In the Bhagavad Gita, Krishna tells us how to become enlightened: "One who neither hates nor desires fruits of activities, liberated from all dualities, easily overcomes material bondage and is completely liberated."[26] Gandhi said, "To see the universal and all-pervading Spirit of Truth face to face one must be able to love the meanest of creation as oneself."[27] The Supersoul, which dwells in the heart of every being, transcends duality. "One who sees the Supersoul in every living being and equal everywhere does not degrade himself by his mind.

Thus he approaches the transcendental destination."[28] With spiritual growth, one sees that there is no evil — it's all good, or all God.

The teachings of Hinduism, meditation, and mantras, as well as Lonna's story, gradually brought me to a turning point around 2002. I understood that if I began to embrace equanimity, as my friend Dana recommended, I could create my own paradise on earth. I had to stop judging. As difficult as that was, Dana mentioned that I eventually looked completely different from her clairvoyant perspective; she could barely recognize me as the person she first met.

As I practiced veterinary medicine, I found peace while being with a suffering animal and holding it in love, just as Dr. Rosenbaum had held Lonna. In daily life, when I witnessed an animal suffering, rather than suffering along with it, I remembered that something equally good was happening. Instead of hating someone for animal abuse, I considered the possibility that I could not punish that person more than his or her own karma would, so I was better off minding my own karma.

Lonna is grateful for her near-death experience. Through her suffering, she gained great insight. Even when we do not understand the mystery behind suffering, we should not grieve, according to Hindu teachings, but rather be free of dualistic judgment and see God in everything that happens.

Ancient Hindu literature references the science of astrology, and an astrologer predicted Lonna's profound experience. The Hindus had vast knowledge of astronomy and gravity before Newton. According to Paramahansa Yogananda, the *Kaushitaki Brahmana* contains passages from 3100 BCE indicating that Hindus were very advanced in astronomy. The *Jyotish*, or body of Vedic astronomical expositions, "contains the scientific lore that kept India at the forefront of all ancient nations....*Brahmagupta*, one of the *Jyotish* works, is an astronomical treatise dealing with such matters as the heliocentric motion of planetary bodies in our solar system, the obliquity of the ecliptic, the earth's daily axial revolution, the presence of fixed stars in the Milky Way, the law of gravitation, and other scientific facts that did not dawn in the Western world until the time of Copernicus and Newton [in the seventeenth century]."[29]

The wisdom of the ancient sages perceived that the microcosm paralleled the macrocosm. Deepak Chopra says that India's tradition of Ayurvedic medicine teaches: "As it is in the microcosm it is in the macrocosm, as is the atom so is the universe, as is the body so is the cosmic body."[30] In other words, inside the living body is an entire universe. The chakras of the body are like the bodies in the heavens, with the first chakra being the earth plane and the third being the sun, and above that the higher planes of existence. Likewise, the cells of the human and animal bodies are conscious beings. One-celled creatures are part of God, and the body is made up of cells. Inside each cell are further worlds; electrons revolving around a nucleus just as the planets revolve around the sun.

Lonna's astrologer warned her about Chiron, which names both an asteroid and a character of Greek mythology. Chiron is a centaur — half-man, half-horse — and this powerful image is the logo/icon for the British Veterinary Association. He symbolizes the wounded healer — a healer and teacher who could not heal himself. Chiron was the teacher and foster father of Asclepius, the half-human, ancient Roman god of medicine and healing. Asclepius's father was Apollo, who had the Dionysian satyr Marsyas skinned alive for challenging his omnipotence. According to Michael W. Fox, Chiron commands us "to question authority and to examine the basis of truth and the consequences of conformity and consensus. The Chiron imperative helps overcome the limitations and potential harmfulness of anthropocentric authoritarianism."[31]

Many brave veterinarians have spent their careers doing just that. One example is Allen Schoen, who early on refused to accept the consensus that animals do not feel pain. In spite of what was traditionally done, he insisted on using analgesia to dehorn cattle, among other painful procedures. He encouraged people to stop treating animals like cars and actually had to argue that animals had "feelings."[32] Thanks to Dr. Schoen and many other compassionate humans, we treat animals more humanely today.

Others in the field of agriculture emphasize the need for humans to overthrow conditioned violent responses to nature. We justify war on nature with pesticides, herbicides, and the slaughter of animal pests, just as ranchers justify killing prairie dogs. Our fear drives us to condone violence to our only home — the earth. In consequence, we are killing ourselves by poisoning our food and water and degrading our divine nature.

Dr. Vandana Shiva lectures for a nonviolent agriculture that does no harm to bees and butterflies and earthworms. She says that although we have created chemicals to destroy "pests," real creativity does not lie in destruction. The other beings on this planet are not our enemies. After all, we are all one and the same — human, beast, and every cell in all plants, and every part of creation.

Still, during the 2002 wildfires, what creative solution would have saved Earl and Janet's fields from the prairie dogs? In the Bhagavad Gita, Krishna told Arjuna to fight to kill his kin because it was his duty, and not to lament their death. Even today, I have a duty to keep Earl's fields healthy and to feed my horses. If I choose to kill, I am responsible for the karma I generate, and I do not want to come back in a prairie dog colony looking up at Dolores Schmidt's shovel. Perhaps, as the ancient Chinese suggested, the prairie dog problem back then was a creation of my own qi. As Jiddu Krishnamurti repeatedly said, "The observer is the observed."[33] Therefore, I was the drought, the fires, and the prairie dogs.

One day that October, as I stared out of my office window pondering this dilemma, I saw a large bird — a bald eagle. She soared without a flap of the wings, moving closer. Then she tucked and dove faster than my eyes could follow through the cottonwoods, landing on a prairie dog in the colony across the road from my house. A passing motorist stopped to witness the kill. Just when I needed an answer, the eagle brought it. She represented the Great Spirit, the connection to the divine, with the ability to live in the realm of the spirit and yet remain connected and balanced within the realm of earth. The One keeps the balance.... It is not my job.

I find the lessons I have learned from spiritual teachings thus far also correlate with what I know from science. In summary of my understanding and how it equates to science, I offer a poem.

THE OBSERVER IS THE OBSERVED

> Jiddu Krishnamurti said,
> The observer is the observed
> The mind creates the illusion of separation,
> Yin and yang,
> The thinker the thought,
> The seeker, that sought,
> Duality.
>
> Hence, one cannot experience Reality
> By thinking.
> The mind must be still.
>
> Prairie dog digs a thousand holes,
> Through the landscape of my soul;

Bald Eagle dives down from heaven,
Skewers the rodent in seconds,
Divine intervention.

Immanuel Kant proposed
A Reality our minds impose.
Experience is an intellectual structure.
When Spirit appeared in the Bisti desert,
Mind formed a white owl from it.

Perceptions influence that perceived,
Interpretation of Reality depends upon one's
Conditioning, anxiety, experience, ambition,
Fear, prejudice, and opinion.
Ancient Chinese called this Qi,
Hence, when Qi flares up, one sees prodigy.

In Werner Heisenberg's estimation
The scientific method alters
The object of investigation.
Human measurements are not objective.
Even our instruments echo the subjective.

Observation without evaluation
Is the highest form of wisdom.
Without dichotomy, the mind is clear
In alert silence, One — Aware.

CHAPTER 6

Buddhism: Finding Peace of Mind

We have names, we have jobs, we have families, and we have all these things that make us think that we are — that we exist as independent entities. And we are always working to establish our safety. Our egos are always working to keep this "I." But in fact, we are just energy systems which are imbued with enlightened energy. And the pattern of ego is overlaid upon that. And so how do we change? How do we return to reality, which is not the "I"? Reality is the actual energy system which is empty, cognizant, and fundamentally compassionate. All Buddhist meditation practice brings us to the experience of that truth, reality, rather than illusion and delusion.

— Lama Tsultrim Allione

Take a deep breath in, and let it out slowly. Watch the next breath come in; notice the air move past the nostrils and feel the chest relax on the exhalation. Sit quietly, watching the breath, and the mind calms as a pond stills after the wind stops blowing. This is a meditation practice used in Buddhism.

The whirlwinds of everyday life — the clocks, the streetlights, the need to succeed — agitate us so that we lose the composure of our true nature. The busy, confused, stressed intellect projects a turbulent and false view of reality. Our true mind is like the pool of water, which reflects best when undisturbed.

Sit in meditation, undistracted by schedules and worry, letting mental activity clear. Watch the breath long enough, and thoughts become less important; self-centeredness dissolves, and one merges into a larger consciousness, becoming aware of the Big Mind that connects us all. Here is a state of being that is free of negativity and afflictive emotions. This is peace of mind and the path to Buddhahood. A Buddha is "one who has awakened to the truth."[1] A Buddha comprehends reality.

The Nature of Buddhism

The historical founder of Buddhism was the man known as Siddhartha Gautama. Born as a prince in India, he gave up wealth and power in search of a way to relieve suffering. He became the Buddha by sitting in meditation under a fig tree in India in the year 530 BCE. There, he realized what modern scientists only recently began to comprehend with Einstein's theory of relativity and quantum physics in the twentieth century. He became "enlightened" to what Buddhists call "emptiness," realizing that no "things" exist separately from anything else. The Buddha said, "Dear friends, I have seen deeply that nothing can be by itself alone, that everything has to inter-be with everything else."[2]

The concept that everything is connected in a basic oneness of the universe is understood in Buddhism and is also one of the most important revelations of modern physics. Physicists tell us that a subatomic particle can only be understood in terms of its activity — its interaction with the surrounding environment. A particle cannot be seen as an isolated entity, but it has to be understood as an integral part of the whole.[3]

The Dalai Lama, the Tibetan Buddhist leader, explains, "Whatever identity we give things is contingent on the interaction between our perception and reality itself. However, this is not to say that things do not exist. Buddhism is not nihilistic. Things do exist, but they do not have an independent, autonomous reality."[4]

To understand this better from a scientific perspective, first realize that the way we perceive reality depends upon the messages our eyes and other senses send to our brains. How one interprets this information is colored by the perceiver's conditioning, fears, and opinions. This explains why twenty witnesses describe the same accident twenty different ways. Not only do different minds understand things in unique ways, we make things up.

Buddhism teaches that all we are is the result of our thoughts. The mind even creates what we call "life" and "death."[5] With our thoughts, we make the world, so we must be mindful of what we project mentally. Buddhism teaches that if we speak or act with harmful thoughts, trouble follows like the cart pulled by the horse. If we speak or act with harmonious thoughts, happiness follows us as our own shadows.

The Buddha realized that the world is "empty," which, from the scientific perspective, describes physical phenomena on the molecular level. Imagine a wooden table. The molecules of the wood are made up of atoms separated by electromagnetic energy, just as the planets are held around the sun. Most of the solar system is empty space. The same goes for the table. Although we see a table, in reality our eyes pick up the light energy reflecting off the electromagnetic energy between the atoms of the table's molecules. The brain translates this information, creating a structure we call the table. Furthermore, the atoms are made up of particles (electrons, neutrons, and protons) separated by energy, which contains empty space. Less than 5 percent of the universe is actual matter. The dining table is a mental construct.

The fact that the Buddha realized this by sitting in meditation is fantastic. He also saw that no "things" exist because everything is constantly changing and interconnected while in the motion of transforming; nothing ever exists long enough for it to have an identity. All things and events perceived by the senses are interrelated in an inseparable web according to both Buddhism and quantum physics (and the teachings of shamanism, paganism, and Hinduism).

All manifestations are part of the same ultimate reality. Our tendency to divide the world into individual and separate things and to experience ourselves as isolated egos is seen as an illusion created by our measuring and categorizing mentality. Fritjof Capra writes, "It is called *avidya*, or ignorance, in Buddhist philosophy and is seen as the state of a disturbed mind, which has to be overcome."[6]

As fascinating as the connection between Buddhism and quantum physics is, the Buddha Gautama did not achieve enlightenment in order to have a great scientific mind. His aim was only to relieve suffering. His goal was compassionate, and his doctrine was not one of metaphysics but of psychotherapy. He did it for peace of mind.

From a religious perspective, Buddhism is based on love and compassion. The Dalai Lama calls his religion "loving-kindness." The divine is not thought of as a ruler who directs the world from above, but a principle that controls everything from within. The pristine, primordial awareness known as the Ultimate Truth, or Buddha-nature, is called "God" by Jews and Christians,

"Brahma" by Hindus, and the "Hidden Essence" by Sufi mystics.[7] Buddhists deny such names, because the Universal Mind is pure in its own nature and free from categories.

Heaven is not a place one visits after death but is right here now, according to Zen Buddhist Thich Nhat Hanh: "There are people who believe that they can enter the kingdom of God, the Pure Land, after they die. I don't agree with them. I know that you don't have to die in order to go into the kingdom of God. In fact, you have to be alive to do so. You should be alive, and you should take one breath in and out, and with one foot you make a step and you enter into the kingdom of God right now."[8]

Although Gautama attained Buddha-nature, he never claimed to be anything more than a man. Instead, he taught that everything and everyone, including an animal, has Buddha-nature, and the truth is within oneself. Upon enlightenment, he said, "I have seen that all beings are endowed with the nature of awakening."[9]

The Dalai Lama introduces Buddhism in terms of two basic principles. The first is the interdependent nature of reality, and because we are all interdependently connected, the second principle is that of nonviolence.

Violence to another human, animal, or the environment is violence to oneself. We are all one connected life. The example being: We are what we eat. The milk we drink comes from a cow that ate grass made from minerals, rain from clouds, and sunlight. Therefore, we are the clouds, the rain, the sunlight that made the blades of grass grow, the minerals in the grass, and the cow herself that gave the milk. If we recognize how interconnected we are, we must be nonviolent and appreciative of the entire whole.

One does not understand Buddhist philosophy from reading about it. Buddhism is an experiential process of knowing, which takes practice. We can, however, study more of the Buddhist teachings about animals.

Animals, Reincarnation, and Karma

The scriptures say that he began his path to enlightenment during
a lifetime when he was a bull in one of the infernal realms. Pulling a cart,
he felt compassion for the weaker animal joined with him. When he told the
demon in charge of this particular hell realm that he would pull the load alone,
the demon became so enraged that he struck him on the head with his trident,

killing him on the spot. Thus the Buddha, by putting another being's needs before his own, began his path to perfect enlightenment and Buddhahood.

— LAMA SURYA DAS[10]

A fundamental teaching of Buddhism is the doctrine of *anatta*, meaning the personification of the self is another illusion. Since nothing is permanent, there can be no permanent soul. Yet Buddhists discuss reincarnation, also called the "transmigration" of the soul. What transmigrates, rather than a permanent soul, are mind states — perceptions and feelings — energetic principles. When the body dies, these energies take some other form.

As with every religious study, there are countless interpretations, and some Buddhists do not believe in the transmigration of souls. However, since Buddhism evolved from Hinduism, numerous descriptions of reincarnation occur, such as the stories of the Buddha's past lives, like the one above.

Six huge volumes of stories known as Jataka, or "birth tales," give the accounts of the Buddha's previous incarnations. They comprise the most ancient collection of folklore in the world.[11] These are life lessons learned when the Buddha was a deer, a partridge, an elephant, a boar, and many other animals and people. He was a bodhisattva, one who is on the path to becoming a Buddha. On the day of his enlightenment, the Buddha is said to have remembered more than a hundred thousand of his past lives.

These stories tell us that we are all Buddhas in the making and that there are animal bodhisattvas, too, such as the llama that died defending his herd from three bears during the 2002 wildfires in La Plata County. Perhaps, through his self-sacrifice, he advanced to an improved incarnation. The Jataka read like fairy tales with morals, such as salvation through self-sacrifice, loyalty, the power of prayer, honor thy mother and father, do not quarrel, the virtue of patience and compassion, and why we should not kill animals.

The story of the "monkey bridge" tells about a king who went into the Himalayas to pick some exquisite mangoes only to find a band of monkeys eating them.[12] The king ordered his men to shoot them and serve their meat with the mangoes. The nearest tree was too far away for the monkeys to jump to freedom; they were trapped. The chief of the monkeys took a huge leap onto the branch of a neighboring tree and made a bridge with his body between the two trees. He told the monkeys to run over his back to safety, which they did. The last one jumped heavily on his back and broke it. The king saw the whole event and came to the dying chief.

The king said, "You made your body a bridge for others to cross. Did you not know that your life would come to an end in doing so?"

"Oh, king," replied the monkey, "I am their chief and their guide. They lived with me in this tree, and I was their father and I loved them. I do not suffer in leaving this world, for I have gained my subjects' freedom. And if my death may be a lesson to you, then I am more than happy. It is not your sword that makes you a king; it is love alone. Rule your people not through power because they are your subjects; nay, rule them through love because they are your children." The king learned his lesson and ruled with love from then on, and all were happy ever after.

Traditionally, all Jataka are told as if the Buddha is telling the story, and each ends with the Buddha saying something like, "I, the Buddha, was that monkey," or whatever or whoever the story is about.

Immediately prior to his last birth as Siddhartha Gautama, the bodhisattva abided in a celestial realm. According to a story told by Ananda Coomaraswamy, when it was time for Gautama to come to earth for one last life, "the deities of ten thousand world-systems assembled together" and told him that it was time to become a Buddha. So he became a beautiful, white elephant and appeared to his mother, Lady Mahā Māyā, in a dream. The white elephant, carrying a white lotus flower in his trunk, came to her, touched her on the right side, and entered her womb. At the moment of her conception, the dumb spoke, the lame walked, the trees flowered, and lotus flowers covered the earth. At his birth, the boy "stood upright, took seven strides and cried, 'I am supreme in the world. This is my last birth, henceforth there shall be no more birth for me!'"[13]

Coomaraswamy refers to this version of the life of Gautama as a "circumstantial biography" rather than a historical account.[14] However, Khenpo Karthar Rinpoche explains why he believes that this teaching is actually true, that the Buddha walked and talked at birth.

When Khenpo Karthar Rinpoche grew up in Tibet, there were no cars or even bicycles, and someone once visited him and told him about airplanes. The idea of flying houses that carried five hundred people was incomprehensible. Since he had never seen anything vaguely similar to an airplane, the story of them seemed unbelievable. Since the rinpoche claims to have seen such things as newborns walking and talking, he believes the story of Buddha's birth. Even so, he encourages us not to believe what others say but to "trust in ourselves."[15]

My interest is in the Buddha taking the form of a white elephant in order to be conceived. Remember, shamanism and other ancient religions teach that a spirit can take the form of an animal. The white elephant image reminds me of similar celestial messengers or spiritual emanations, such as the white buffalo for Native Americans and angels in Christianity. The appearance of the white elephant in the dream symbolizes the birth of an extraordinary person, Siddhartha.[16]

The Buddha did not discuss reincarnation. He taught for forty years, saying over and over again, "I teach only suffering and the transformation of suffering."[17] He focused on happiness in this life, in the present moment. "My teaching does not depend on whether I exist after death or not because I am concerned with suffering here and now. Let the past be and forget the future. I will teach you that which is now."[18] Although the Buddha did not discuss life after death, he accepted the Indian beliefs about karma and rebirth and about deliverance from rebirth by reaching sainthood (nirvana).[19] Brahmanical and Buddhist thought remained tightly enmeshed.[20]

The source of Buddhist teachings began with the Buddha's enlightenment experience when he gained direct knowledge of the central teachings of early Buddhism: karma, reincarnation, and the Four Noble Truths.[21]

The Buddhist view of cosmology is that the universe has no beginning and no creator, only countless world-systems with no known limit, which come from countless "big bangs."[22] Within this vast universe, beings go through endless cycles of rebirth into six realms of samsara — the cycle of conditioned existence: gods, demigods, humans, animals, hungry ghosts, and hells. Each is the result of six main negative emotions, respectively: pride, jealousy, desire, ignorance, greed, and anger.[23] Hence, animals are said to live in the mode or realm of ignorance, and humans in the mode or realm of desire.

The animal existence includes sentient creatures as simple as insects.[24] Plants are not included, though they are seen as having rudimentary consciousness in the form of sensitivity to touch. Some realms of beings are not (normally) visible, such as the realm of *petas*, the "departed," ghosts with bodies made of subtle matter. Such a rebirth does not involve reincarnation into a gross physical body. *Petas* are seen as frustrated, hungry ghostly beings that frequent the human world due to their strong earthly attachments.

The worst realm is the hell realm (*niraya*), involving experiences of being burnt up, cut up, frozen, or eaten alive. The animal, *peta*, and hell realms are the lower rebirths, where beings suffer more than human beings. The higher, more fortunate realms of rebirth are those of humans, devas, and gods. The gods are just enjoying the fruits of good karma.

Karma determines the realm of rebirth and is the psychological impulse behind an action; karma is action with intention. Buddhism and Hinduism teach that animals live in the mode of ignorance; they lack intention behind their actions and are not responsible for their behavior with regards to karma. In other words, a cat can kill a mouse without accruing bad karma. Cambridge professor Peter Harvey explains: "Animals, ghosts, and hell-beings have little freedom for intentional good or bad actions, though the higher animals can sometimes act virtuously, if not in a self-consciously moral way."[25] This virtuous conduct is what the Jataka tales describe as the bodhisattva path. However, for most animals the mode of ignorance keeps them from being responsible for their actions; karma is mostly important in human incarnations.

Professor Harvey explains how karma affects reincarnation: Beings in the lower rebirths reap the results of previous bad actions. When these results come to an end, the results of some previous good actions will come to fruition and buoy up the being to some better form of life, sooner or later reaching the human level.

A soul is likened to a stream or the flame of a candle. In rebirth, the flame moves from one candle to another. Enlightenment or nirvana means cessation.[26] With nirvana, the flame (of suffering) is blown out. Nirvana is not just the cessation of life and reincarnation; it is a quality of mind or state of being that characterized the Buddha's life in the forty years between his awakening and his death, when the fire of his personality finally flickered out.

Samsara and nirvana are not external things; they are internal. No one is created in one form or another; we ourselves are the creators. When we remove the causes of suffering — the negative emotions of greed, hatred, ignorance, attachment, pride, and jealousy — we reveal the positive qualities beneath them: wisdom, awareness, loving-kindness, and compassion. Then we are liberated from samsara, but we are still in this world.

The distinction between an enlightened existence and unenlightened existence is based on levels of consciousness, according to the Dalai Lama: "A person whose mind is undisciplined and untamed is in a state of samsara or suffering;

whereas someone whose mind is disciplined and tamed is in the state of nirvana, or ultimate peace."[27]

I sometimes hear Buddhists say that animals are not "conscious," which from a veterinary standpoint is nonsense. An animal is either conscious, asleep, under anesthesia, in a coma, or dead. I have long struggled with the same notion presented by scientists who believe that animals are not conscious. Consciousness is a difficult word to define. The word *conscious*, according to the Dalai Lama, refers to the level of control a being has over their mind. Animals run more on instinct than mental discipline. They do not consciously attempt to control their thoughts; therefore, they are less "conscious." I suspect there may be a Tibetan word that better defines one's level of mind control that the English language lacks, and we only have the one word, *conscious*, which does not fully express the meaning the Dalai Lama intends.

However, let's consider a cat stalking prey. The cat demonstrates focused concentration and controls the urge to attack until the time is right. Some animals use trickery to catch prey or to avoid a predator. They seem to be very aware and in control. Whether they analyze their thoughts and mentally control themselves by thinking, *Wait, not yet*, we cannot know. I suspect they follow their inner guidance, the Buddha-nature that tells them, *Go now, run*. We call their inner knowing *instinct*, and we believe that instincts do not require conscious thoughts. Indeed, instincts work best when we do not think, as I experienced with karate, when my body defended itself without thought. I believe my reactions came from my "spirit," or internal guidance, or Buddha-nature, rather than mental control. All beings have Buddha-nature. Furthermore, the Dalai Lama does not say animals are not conscious, he says they "are less conscious" and that even plants have rudimentary consciousness. It seems more likely that varying degrees or levels of consciousness exist among various beings and even at various times. I occasionally find myself driving, lost in thought, then I wake up to wonder where I am going. With regards to degrees of consciousness, we may even know some animals that seem more conscious than some humans.

According to Tibetan Buddhism, animals cannot become enlightened in their animal incarnation. Only in the "precious human body" is enlightenment possible. In chapter 5, I mention that the mystic Sadhguru said that it is possible for an animal to become enlightened, but it is extraordinarily rare. I have heard tales of such animals. In addition, Tibetan Buddhism evolved from and included some teachings of shamanism, which teaches that a spiritual being can take the form of an animal. Therefore, it seems plausible that an enlightened being could appear in the form of an animal at times. But because it is extremely rare for a human being to become enlightened, most Buddhist teachers laugh at the idea of an animal reaching nirvana in this lifetime.

The animal realm is supposed to have more suffering than the human realm; animals are killed for food and if they present a nuisance. However, the average American pet appears to enjoy the life of a demigod. Let us examine what the Buddha says about suffering and how it affects animals.

How to End Suffering

The Four Noble Truths are the main teachings of the Buddha about suffering and how to end it. To relieve suffering, we must first realize that suffering exists — that is truth number one. There are three kinds of suffering.[28] The first is the physical kind associated with pain from trauma, sickness, and aging (*dukkha-dukkha*). Both humans and animals suffer physical pain.

The second kind of suffering is due to change (*viparinama-dukkha*). Everything decays or falls apart; everything that's joined gets separated. We all suffer from the loss of loved ones and from losing things such as cars. Animals do not cling to objects as desperately as we do. However, a horse will run the fence for hours wanting to follow a companion that left the pasture; a dog will destroy a house over "separation anxiety"; and a wolf will howl for days over the loss of its mate. Animals also experience the suffering caused by change.

The third kind of suffering is the suffering due to conditioned states (*samkhara-dukkha*). Perhaps humans suffer more from this kind than animals do because, for example, as soon as we get a car, we worry about it being scratched, stolen, or dirtied. This is how a pleasurable thing can cause pain in the midst of pleasure. We project worry into the future. House pets become anxious when their people pack a suitcase, and horses get excited when a horse trailer arrives. People worry themselves sick, and domesticated animals sometimes do the same. Wild beasts become so frightened when captured that they may die from the stress. Although animals face concerns about possible future scenarios, such as the fear of prey, they seem less likely to worry for days about something they did or what may happen next week, as humans do. Animals appear to be more present in the now.

Thich Nhat Hanh suggests that we should behave more like wild animals when we have health problems. "When animals in the forest get wounded, they find a place to lie down, and they rest completely for many days. They don't think about food or anything else. They just rest, and they get the healing they need. When humans get sick, we just worry! We look for doctors and medicine, but we don't stop.... Just allow your body and mind to rest like an animal in the forest. Don't struggle."[29]

The Second Noble Truth describes the origin of suffering. We must realize that we create our own suffering by worrying and struggling over things and events that are impermanent, ever-changing mental constructs. Animals provide a good example of how to release attachments more easily. In the *Dhammapada*, the Buddha describes freedom from desire and sorrow: "Like a swan that rises from the lake, with his thoughts at peace he moves onward, never looking back."[30]

The Third Noble Truth describes the way to end suffering. We must change our minds to rid ourselves of delusions. Buddhist meditation practices provide the way to end suffering. The Buddhas and bodhisattvas know how to transform suffering into compassion and joy, while the average human struggles in a conditioned existence.[31] We project, worry, and agonize over past events. With meditation, we control the mind enough to stop focusing on these troublesome thoughts.

The Fourth Noble Truth teaches the path to the "Eight Right Practices," also known as the "Eightfold Path," that lead us away from suffering: right understanding, right thought, right speech, right action, right livelihood, right effort, right mindfulness, and right concentration. Buddhism teaches that animals do not realize mentally that they create suffering, nor that they can stop it, therefore they have no way to escape it via enlightenment, as we do in these "precious human bodies." I must point out that most humans do not escape suffering in this manner either. Meditation practice helps me.

Animal Sacrifice and the First Precept

The Buddha called for the end of animal sacrifice. He said, "Greater than the massacring of bullocks is the sacrifice of self. He who offers up his evil desires will see the uselessness of slaughtering animals at the altar. Blood has no power to cleanse, but the giving up of harmful actions will make the heart whole."[32]

A friend once told me, "The thing I like most about Buddhists is that they don't kill anybody." In fact, that is the first precept, a vow one makes when becoming a Buddhist. The first precept is regarded as the most important. It is the resolution to not kill or injure any human, animal, bird, fish, or insect.[33] Historically, this did not mean that Buddhists were vegetarian, and the Buddha accepted meat in his alms bowl. The emphasis was on avoiding intentional killing, so it was worse to swat a fly than to eat meat.

The Buddha allowed a monk to eat flesh if the monk had not seen, heard, or suspected that the creature had been killed specifically for him. Such food was called "blameless." Even killing to give meat as alms generated "demerit" (bad karma), due to the distress felt by the animals while being brought to slaughter and the pain when killed.[34]

Various sects of Buddhism interpret meat consumption differently. Some deny that the Buddha allowed "blameless" meat for monks. Buddhists also give these reasons to stop eating meat: All beings have been relatives in a past life; eating meat hinders meditation and leads to bad health, arrogance, and rebirth as a carnivorous animal or low-class human; and if no meat is eaten, killing for consumption will cease.[35]

Buddhist sects differ in their views regarding vegetarianism. Many are vegetarian, some Japanese Buddhists eat fish, and in the colder climates (like Tibet), vegetarianism is impractical. Even so, Buddhists show deep compassion to other sentient beings. Some avoid even eating honey, for this is seen as theft from and murder of bees.

The first precept not to kill creates a moral dilemma for the practicing Buddhist with pets at the end of their lives. Recall Margaret, the Buddhist woman, and her dog, Jaws, from chapter 1. Margaret felt obligated to let Jaws suffer his karma in this life so he would have a better incarnation in the next, and at the same time, she did not want the dog to suffer, and she brought the problem to me to solve. I now know several important things I did not know then that pertain to the first precept. First, as I describe above, karma is not as important for animals. Indeed, the primary purpose of the first precept is not to prevent animal suffering or protect an animal's karma, but rather to protect humans from karma. Even in Jainism this is true. Jainism has strict rules about killing, yet avoids sentimentalizing animals. Hence, Jains do not own pets because pets are carnivores and also have parasites, so one might have to kill on their behalf. "Ultimately, the reason one respects animals is not for the sake of the animal, but for the purpose of lightening the karmic burden that obscures the splendor of one's own soul."[36]

The second important thing I have learned is that the Buddha taught the end of suffering and said that the first precept can be broken for the purpose of relieving suffering.[37] Of further importance is listening to our own Buddha-nature when making challenging decisions. In Theravada Buddhism, the discourse from the *Tripitaka* contains some teachings the Buddha Gautama passed on to his disciple, Ananda. Here the Buddha repeats the lesson to trust the truth within. He told Ananda: "Take refuge in nothing outside yourselves. Hold firm to the truth as a lamp and refuge, and do not look for refuge to anything besides yourselves."[38] I may not be as wise as Ananda, but this teaching is so replete among high-level spiritual beings that I now trust my inner guidance about when animal suffering permits euthanasia. Each person must make their own decisions in this matter rather than following dharma, dogma, or the advice of other people.

The biggest fear for people regarding veterinary matters is the fear of making some wrong choice that causes an animal to suffer, especially at the end of a pet's life. After questioning numerous Buddhist teachers on euthanasia of pets, I return to the Buddha's teachings. The intention behind the action is what's important. Each person must trust the truth that is within and act to relieve suffering.

Three stages of karma determine whether it is complete or incomplete: intention, action, and satisfaction. When intention motivates the action and a sense of satisfaction follows, then the action is complete and generates karmic consequences. In other words, if one kills an animal by accident, without intention, the consequences are less severe; without the feeling of satisfaction, the action also remains incomplete.

In my job, rather than being concerned about the animal's karma in the next life, I focus on relieving suffering now. Humans suffer more anxiety, fear, guilt, and worry if they are told to keep an animal alive who is in agony and has become a burden for the family to care for. If a seventy-year-old woman cannot lift her sixty-pound dog that is unable to stand, the suffering must end. Perhaps it is their good karma that I show up to put an end to the grief. My conclusion is this: The person who loves the pet most must make the choice and live with it; I am there to help them and the animal. My intention is to relieve suffering; thus I act, and although the process is unpleasant, people feel relief when it's done.

Veterinarians must do their jobs to serve patients and clients the best they can under difficult circumstances.

A veterinarian has to kill. This includes more than animals at the end of their lives. It also pertains to countless parasitic insects, such as intestinal strongyle parasites that kill horses; I must kill the worms to save the horses. For veterinarians, these killings are our duty, and they provide deep insights of compassion for human and animal alike. Deep contemplation of suffering and killing led me to Buddhism.

Chocolate Nirvana versus Tonglen

I stood at the checkout counter of my local natural foods store feeling embarrassed as I became aware of my purchases. They included chocolate yogurt, chocolate soy ice cream, frozen vanilla rice cream, hot fudge habanero sauce, chocolate graham crackers, chocolate almond milk, an avocado, two bananas, and four bars of organic dark chocolate.

"I love chocolate, too," said the young lady at the register. There before me was a reflection of my poor mental and emotional health. In 2003, a summer of emergency calls had left me distraught. I had given up alcohol to practice Buddhism, so chocolate was my last attempt at escape. Buddhists take refuge in the Buddha, the dharma (Buddha's teachings), and the sangha (the Buddhist community); I was hoping to find nirvana in chocolate. All was suffering, as the Buddha said, and that was all I saw. Not only was I exhausted physically from working long hours, my emotions were tortured by visions of dying animals replaying in my mind like images from horror movies.

Three horses I had treated for many years died of dehydration because the gate to their water source became "accidentally" locked during the unusually hot, dry July that year. The owner was out of town, and a neighbor called me at dusk when he found one horse dying at the gate. We searched the pasture in the dark on a four-wheeler for the others. We found Knight, a once-majestic, black quarter horse, dead at the bottom of a dried-up pond. I stood on the rim, shining a beam of light from my headlamp onto his body. Dust particles in the air obscured the view. He was curled up and looked like a desiccated fetus in a bowl of gray powder. I imagined him pawing for one drop of water in the hot sun before he died. I felt weak and nauseous as we searched for the others. We found another dead mare and two more horses still alive and out of their minds from shock, a yearling and a bay gelding whose abdomen was so drawn up I could not identify him as a male at first.

Two nights later I was up with an old, gray horse in agonizing pain. I still heard the sounds of little girls screaming in horror as they ran into the bushes to hide while I euthanized their friend.

If that was not enough, the desperation in a beautiful red fox trotter's eyes haunted me, and I could not shake the sight of the horse throwing his body down again and again, thrashing in the gravel from colic pain during a violent thunderstorm, until I finally gave him peace by ending his misery with a lethal injection.

On top of those horrible events, one client was blaming me for his horse's ailments, unable to accept responsibility for his own negligence, and refusing to pay me, and another was accusing me of causing his horse's accidental death. The man had tied his horse to the bumper of his truck with a long rope, which is always dangerous for the horse. (Please, always tie a horse to something head high and on a short lead.) I did not notice what he had done because I was figuring his bill on the computer at the back of my truck. Then the horse walked briskly next to me and hit the end of the rope between her legs, which jerked her chin to her chest; she fell to the ground and landed on her head, breaking her neck. She died instantly at my feet. I did not at the time blame the man for his ignorance, but afterward he still did not understand his error and blamed my vitamin B12 injection for killing her.

Trouble was following me like the cart behind the horse, and I felt helpless to control any of it. Sure, I had reasons to eat chocolate, but I needed to learn to handle the suffering. Either I would have a nervous breakdown or die from chocolate toxicity, or I would have to change.

Over the next year, I increasingly looked to Buddhism for help, not only to understand animals better but reality as well. I needed more joy in my life. I sat in meditation and read a great deal. Then one day I asked the universe to send me a connection to a Buddhist teacher. Three days later, Veronica Wilson called with questions about her sick cat. Although she lived hours away, she called me because, as synchronicity would have it, her friend knew me; he was from my hometown in Wisconsin. Veronica worked at a Buddhist retreat center called Tara Mandala, run by a Tibetan Buddhist named Tsultrim Allione, and Veronica invited me to hear Tsultrim speak in Durango that month — March 2004.

Tsultrim's lecture and guided meditation was on the Tonglen, a practice in compassion. We gathered in a third-floor room of an old junior high school. A stack of wool blankets, pillows, and zafus (meditation cushions) sat at the entrance. There also was a small stand supporting a basket for donations. I put some bills in the basket and took my seat near the center of the room on a green pillow over a pink wool blanket.

As I settled in, I recognized six of my clients, people I never knew practiced Buddhism. One was Rachael Thompson, a woman I had known for fifteen years. We hugged, and she sat next to me. She glowed. Her white hair radiated in all directions like an aura. Her smile was contagious. She told me about a Tibetan lama she had met while she and her son were traveling in Mexico. They immediately fell in love with him and planned to bring him to Durango to offer a retreat. Of course, I signed up right away.

Birds were singing in the branches of the trees outside the open third-story windows. I felt like we were sitting in the trees with them. Tsultrim surprised me because she was so beautiful, with long gray hair and wearing jewelry and lipstick. I expected a plain, bald nun. Rachael explained that Tsultrim, one of the first Americans to become an ordained Buddhist nun, had renounced her Buddhist vows to be married and have children. To begin, we sat in meditation, and then Tsultrim started a discourse on Tonglen. She explained that the purpose of Tibetan spiritual practice is to benefit all sentient beings.

Sentient beings are defined as those beings that move, or *droas*. Plants move, but not fast enough to be considered sentient. Tibetan Buddhists believe there is no liberation for anyone until all beings are enlightened, so they practice to benefit the whole. People generally practice meditation for stress relief or because it feels good, but Tsultrim taught that we must practice for other beings because developing loving-kindness for others is what brings us happiness.

Tonglen means breathing in the suffering of others and breathing out love and compassion. The person meditating takes on no ailments in this process because the practice takes place within the framework of emptiness; no one can dump anything on another because that reflects the duality of self and other, which does not exist in emptiness. Once you make your heart larger, you take nothing on. The practice is done in the great clear light of "absolute bodhicitta" — the true nature of mind, separate from mental fabrication. This mind is vast and luminous, free of dualistic subject-object cognition.

In Tsultrim's presence, I easily slipped away into the clear light. When she askcd for questions, I asked if the light was the same as the light and feeling of surrender experienced by people who have near-death experiences.

Tsultrim explained, "At the time of death, the fabric of the body breaks open and the consciousness leaves the body. At that moment, there is an experience of what they call 'the child leaping into the mother's lap.' The mother is Prajna Paramita — called 'the Great Mother.' You go into the lap of the nature of mind. Everyone has a flash of this at the time of death. If you practice, you can be trained to recognize it; most people miss it. There is that openness for a moment. If you recognize it and go into it, you can be liberated. At the moment of death, all there is is the nature of mind. That is why the Tibetan teachings put so much emphasis on the moment of death. There is that opening before you go into the bardo — the process of moving to the next life and starting the whole mechanization over again."[39]

Tibetan Buddhism also describes the mirror that Lonna, the woman who had the near-death experience after being kicked by the horse, saw in the emergency room. The Avantansaka Sutra states: "The one true essence is like a bright mirror, which is the basis of all phenomena, the basis itself is permanent and true, the phenomena are evanescent and unreal; as the mirror, however, is capable of reflecting images, so the true essence embraces all phenomena and all things exist in and by it."[40]

Tonglen is first practiced for our mothers. Mother is a metaphor for compassion because mothers, even animal mothers, manifest compassion. They go beyond themselves to care for their young; they break through their limits and their self-orientation. Our mothers give us these precious human bodies, which are vehicles for enlightenment, something to be very grateful for.

Buddhists believe that for a person to discover these teachings in this lifetime is unusually exceptional. Tsultrim repeated the classic story in which the Buddha described how rare it is to incarnate into a precious human body, one that is well-favored enough to encounter these teachings and understand them. The Buddha told a story that captures this: Imagine that all the earth were covered with water and a man threw a yoke (a collar for a plow) into the water; the odds of being born into a human body that can attain enlightenment are so small that it is like the chance that a blind sea turtle, which only comes up for air once every one hundred years, would come up into the center of that yoke. Since nirvana is rarely reached by humans, I now understand why Buddhist teachers laughed when I asked about animals becoming enlightened.

One of the hardest parts about suffering is watching it. We do not want to be present with suffering, so we turn away, sometimes by using addictive substances. Instead, if we bear witness, and practice Tonglen, breathing in the pain and sending out compassion, we can be bodhisattvas and we can change the world. We also need to feel compassion for ourselves. You and I are not separate; we have each other's pain. In realizing this, we see that emptiness gives birth to compassion, and compassion gives birth to emptiness as well.

On the subject of the enlightened mind, I asked Tsultrim, "If animals are not self-aware, and have a nondual, nonmoral way of looking at the world, and since they view life from the present moment, how is that different from an enlightened state?"

Tsultrim replied, "Not being self-aware is not necessarily an enlightened quality. The difference is in the quality of knowing. All sentient beings have the ability to know. Animals have an incredible ability to know about things like migration and survival. However, their knowing operates at the service of passion, aggression, and ignorance. When knowing looks at knowing, when mind looks at mind, it discovers timeless wisdom. The knowing underneath the manipulations of ego is the primordial wisdom uncovered with practice. This is something that only humans can know, and is different from the instinctual knowing of an animal."[41]

I also interviewed Tsultrim about the possibility of animals meditating. She agreed that animals may look like they are meditating in their simple way of being and not fabricating mentally, but they are not becoming enlightened any more than chocolate will create nirvana for me. Still, Buddha-nature is in all sentient beings waiting to be awakened.

Bear in mind, I am still trying to understand these teachings. I am on a quest to understand things beyond my knowing. I may be as blind as the proverbial sea turtle, and even a glimpse of nirvana would be a fabulous feat for me. I humbled myself to Tsultrim's teachings while connecting with my guidance within. As we sat in meditation, Tsultrim interrupted, telling us to direct our minds to "look now" at our minds. When I did that, I fell into bright bliss. "When mind looks at mind, you see there is nothing there." That glimpse of emptiness showed me how little I knew.

According to Lama Surya Das, "Tibetan Buddhism says that at the heart of you, me, every single person, and all creatures great and small, is an inner radiance that reflects our essential nature, which is always utterly positive. Tibetans refer to this inner light as pure radiance or innate luminosity.... There is nothing after this and nothing before this."[42] I glimpsed the innate luminosity, and it felt wonderful. With much practice at meditation, this bliss state has become a more frequent experience for me.

Tsultrim Allione teaches that all our demons — inside and out — come from a tendency to focus everything on ourselves. We think about the whole world in terms of what it will bring us or what we are afraid of — what we like and do not like. Actions oriented on that premise are the opposite of compassion. The primordial demon comes from the splitting of self and other. Tibetan Buddhism offers ways to transform fear and other demons through meditation and visualization using mantras. When I practice Tonglen, and other meditations, I free my mind, which is projecting a holographic movie that I call "my reality, my suffering."

Tibetan Buddhism

The most traditional form of Buddhism in existence today is the Theravada ("Doctrine of the Elders"), which was founded in Sri Lanka, Burma, and Thailand. This school of Buddhism holds that they have the earliest surviving record of the Buddha. The Mahayana sect, however, claims to have the teachings of the Buddha himself, which were passed on secretly and concealed for centuries until the world was ready to receive them. These teachings spread to China, Tibet, Japan, Korea, and Vietnam. Two familiar forms of Buddhism in the United States today, from the Mahayana sect, are Tibetan Buddhism and Zen Buddhism. Many people also practice Theravada Buddhism, and in addition, there is a new American Buddhism, which is somewhat of a potluck supper of the other types of Buddhism served with apple pie.

Because Tibetan Buddhists believe in the reincarnation of animals, and Theravada Buddhists do not, Tibetan is most interesting for me. The familiar images of Tibetan Buddhism are saffron-robed monks chanting guttural mantras, the snow-topped Himalayan mountains, and the smiling face of the Dalai Lama teaching loving-kindness. The fourteenth and current Dalai Lama, Tenzin Gyatso, is an incarnation of the Avalokiteśvara — the Bodhisattva of Compassion, here to help humankind along the path to enlightenment.

The Tibetan geography formed forty million years ago when the Indian landmass collided with Asia, forming the world's highest mountains — the

Himalayas — and a huge high-altitude plateau where Tibet is today. When the land was newly formed, so the story goes, the only inhabitants were a monkey and an ogress (or goddess, depending on the version). The monkey was an incarnation of Avalokitesvara — yes, the same Bodhisattva of Compassion who is now the Dalai Lama. He was peaceful and contemplative. The ogress, a creature of wild emotion and lust, driven by sexual desire, wailed piteously for a mate. When the monkey heard her cries, he was filled with compassion, and the two mated. Their union produced six offspring, born without tails and walking upright. These were the progenitors of the Tibetan people. Avalokitesvara continued throughout the ages, reincarnating as many beings, including a king and fourteen Dalai Lamas, and he works to aid in the spiritual development of Tibet.

Tibetan Buddhist religious history is full of colorful characters, such as nagas — semi-divine, dragon-like sea dwellers the Buddha entrusted with the Scriptures of Transcendental Wisdom. The Indian philosopher-sage Nagarjuna discovered them in the first century BCE and brought the highly treasured teachings about emptiness back from the nagas.

The original Tibetan religion was called Bön, a collection of shamanistic and animistic practices. Buddhism spread into Tibet at the request of religious King Trisong Detsen in the seventh century. The proponents of Buddhism incorporated the demons of the indigenous religion while in the process of conquering the Bön priests. Thus Bön and Buddhism entangled. Demonic art is common in Tibetan religion. Tsultrim Allione teaches that these demonic or wrathful forms represent elements of human consciousness that are transmuted through meditation into virtuous qualities. She describes these practices as "cutting through fear." The transformative meditation techniques derived from another source of shamanism — Tantra.

Tantra refers to the "warp" that is used to weave a piece of fabric, the thread on which reality is woven; it also means "power" and a lineage of transmission of power. It is a system of practices and meditations derived from esoteric texts emphasizing cognitive transformation of passion, aggression, ignorance, pride, jealousy, and ambition through visualization, symbols, and ritual.

The Tantric meditative practices came into Tibet with Padmasambhava, a Tantric yogi and saint, also known as Guru Rinpoche. He was a sorcerer, a Tantric magician who answered King Detsen's call to bring the Buddha's teachings to Tibet. With superior magic, he subdued the Bön.

Padmasambhava first appeared in a giant, multicolored lotus flower that floated in the ocean. A king saw a multitude of birds flying over the lotus and went to see what was there, only to find an eight-year-old boy in the lotus flower who claimed to be the lotus-born Buddha.

Padmasambhava could also walk on water. One day he needed water to make tea, so he hit his staff on a rock, and a stream came forth that still runs today. Once, when there was an attempt to burn him alive on a pyre, he turned the fire into a lake, which is also still present today. He is considered to be the father of the Tibetan religion, "the second Buddha," the carrier of Tantric Buddhism into Tibet in the seventh century.[43]

Tantra is found in Hinduism and Buddhism with vague origins and secret transmissions from guru to disciple. Mantras are also important in Tantric practice. According to Lama Tsultrim Allione, the roots of Tantra are from the ancient goddess religion.[44]

As Tsultrim explains, Tantra arose from very ancient pre-Aryan, pre-patriarchal roots. The Tantric practitioners were ordinary laypeople in the community — parents, merchants, craftspeople — who turned their work into practice. Especially involved were the lower-caste, darker-skinned people. They were the Indian peoples conquered by the Aryans; they lost the war and were dominated, so they took their spirituality underground. Their strong religious roots were in goddess worshipping, and women carried the Tantra.

At the time when Tantra and Buddhism met, Buddhist monks in India were very astute intellectuals running universities. They were involved with politics and were receiving financial gifts. They lived a dry, scholarly existence. The Tantric women started teaching these intellectual monks "in a very direct, juicy way," Tsultrim says. The introduction of Tantra "was an infusion of juice, and of the feminine," that came in with the first women teachers from the lower caste who practiced the goddess religion.

Historians referred to these women as consorts or sexual partners for the yogis, but on closer examination, they were "lineage holders." The monks were rigid with vows of what was clean and unclean. Tantra promoted the idea of going beyond dualism, cutting through the notions of pure and impure, of self and other. Their structures were broken open to embrace all of life and all experience as sacred. Numerous stories tell of monks who sought teachers and encountered women who passed the Tantra on to them. The story of Maitrigupta (also known as Maitripa) is an example of how going beyond dualism offers enlightenment.[45]

As the story goes, Maitrigupta was studying for exams in a great Buddhist monastery of northern India when a dakini, a Tantric messenger, flew through the window and gave him a glimpse of enlightenment, shattering his perception of the world. She told him to go to southern India to find a teacher by the name of "Mountain Man." Maitrigupta went to a remote area to find this guru, Mountain Man, who told him to go out into the forest, find a flat rock to sit on, and not eat for seven days, until he received a revelation. Near the end of the seven days, out from the woods came a wild huntress, carrying a bow, arrow, and knife and chasing a wild pig. She shot the pig and sliced off a hunk of meat from its haunch. She held it out to Maitrigupta, saying, "Here, Maitrigupta, eat this. Eat the flesh. It is emptiness. Taste the blood; it is the great bliss." Maitrigupta went on to become a great teacher of Tantric Buddhism.

The Tantric tradition is about direct experience, encountering reality face to face. Insight comes by overturning conventional distinctions. For a Buddhist monk, nothing is more impure than someone who kills to eat meat. Tsultrim stresses the important lesson: in religious practices, such as the world of the monastery where things are seen as pure and impure, such as bloody pieces of meat and the people who kill animals, seeing things in this dualistic way misses the point. Emptiness cuts through all divisiveness. Tantra teaches that dualism is the cause of all our suffering. It takes a radical overturning of conventional reality to find a union of opposites. In emptiness, there is no pure or impure, and all can transform.

Veterinary medicine certainly provides opportunities to encounter reality face to face. It is indeed "juicy." That year, I began to accept that, if I saw veterinary practice as perfect, without the dualistic thinking of good and bad, and did my duty with compassion, I could be free of much suffering and help others heal. In addition, I started to recognize that when I judged something, or someone, I felt pain and the person judged me. A lifetime habit of dualistic thinking resisted change. I hoped a Tibetan lama could help me.

Dragons and Demons

In late June 2004, three months after meeting Tsultrim Allione, I first met Lama Lhanang Rinpoche in the same third-story room. Jean-Luc and I arrived ten minutes late to the evening lecture and found two spaces remaining in the center of the floor. Our friend Debbie waved us over to join her, and we wiggled into tolerable meditation postures.

Lama Lhanang sat directly in front of me passing his fingers along his mala, or rosary of wooden beads, as he mumbled a mantra. He wore a sunflower yellow shirt and a burgundy shawl. Behind him and all around the room hung pieces of his artwork, drawings and paintings of the Buddha, the Mother of the Buddhas (Tara), Maitreya (the next Buddha to come), Sarasvati on a duck, and pen drawings of mandalas in silver, gold, green, and blue on black paper. He spoke just louder than a whisper about loving compassion. He explained that he used to hunt and fish, but then he decided not to use his body to harm living beings anymore. As he spoke, I saw a lavender and cobalt blue light flash up from his shoulders. That was only the second time I had seen an aura, and it impressed me. I looked forward to our retreat at Blue Lake Ranch planned for the next month.

Six people, plus Lama Lhanang, attended the private retreat at Blue Lake Ranch in the quiet of the country. The flower gardens were in full bloom with rose, peony, and delphinium. Along the lakeshore, iridescent blue dragonflies mated. Rainbow trout, a foot long, searched the weeds as minnows darted for cover. Turtles poked their heads above the water to watch us, and hawks screeched in the clear sky above. Fat does with fawns grazed in the tall grasses, and when a snake slithered across our trail, Lama said, "Ah Ho!"

I was reminded of the Mayan-Hopi-Tibetan connection that fascinated Rachael, who brought Lama Lhanang to Durango. This Tibetan man resembled a Native American, especially when he ended his mantras with "Ah Ho," just as our indigenous tribesmen do. If he was dressed in a cowboy hat and jeans, he could easily pass as a Hopi. Not only do these peoples resemble each other physically but their religious philosophies are similar: They believe that all life is connected, they understand the powers of transformation, and they know there is another life after this one.

I was curious about Lama Lhanang's previous incarnation as one of Padmasambhava's original twenty-five disciples. He was the incarnation of Ken Rinpoche Damcho, an emanation of Numchan Namke Nyingpo — who could "mount sunbeams."[46] I wondered if he still had that ability. He said that he did not have any magical powers, but his teacher did; he healed people, he walked on water, and while he was captive in a Chinese prison for twenty years, he was often seen having lunch with people in Tibet.

With the practice of anu yoga, yogis, like his teacher, went to the Pure Realm and returned to tell many stories. By doing the anu yoga breathing exercises, they learned to fly to other worlds. They kept their body heat strong and stayed warm all winter in Himalayan mountain caves, and they healed by placing hands on diseased areas.

Lama Lhanang was recognized as this important incarnation at the age of nine, and he was taken from his family (and crying mother) to a monastery. He studied the dharma for twenty years in the monastery and then came to Los Angeles three years before coming to Durango. I was struck by his child-like innocence.

Two weeks earlier, Lama Lhanang made it possible for Rachael and her son to see the Dalai Lama in Los Angeles. Lama Lhanang told how wonderful it was to be in the Dalai Lama's presence: "A monk in Tibet has one wish, that someday he might see the Dalai Lama. I was very excited to see him. I had all these questions. Yet when I saw him, I just cried. I said to myself, 'What are you doing? This is the Dalai Lama. Say something.' But all I could do was cry. He held my head in his lap and I cried."

Each day of the retreat we practiced anu yoga and sang mantras. Some afternoons we practiced art. One day Lama Lhanang taught us to draw dragons. He told us that dragons still exist in Tibet; they cause thunderstorms, appearing like purple tornadoes that go up from the ground to the sky. Once Lama Lhanang's father touched a moth, and it turned into a dragon, starting a whirlwind followed by a storm.

He explained that dragons are from the spirit realm, although sometimes they do appear as snakes or insects. Other beings from the spirit realm manifest as animals as well. For example, he said, Vajrasattva — a manifestation of the primordial Buddha, Samantabhadra — appeared as a duck that then floated across a lake to tap the heart area on the chest of a princess three times. She was filled with joy and conceived Prahevajra.

Some spirits manifest as animals, and so do some demons. "You never know which you may be seeing with your outer eyes. You have to use your wisdom eyes to know the difference." A dakini is analogous to an angel, Lama Lhanang said. They are drawn as naked women with a necklace of skulls, but they have the same function as angels, acting as messengers. Tsultrim Allione said dakinis are the female embodiments of enlightened energy, and a dakini can appear as an animal. They have an enlightened being language and can also speak animal languages.

Animals can see "the reflection world" and can communicate with the spirits. That is how they sometimes know things that we do not, and why shamans can talk to the animals to receive information about the weather, for example. Animals can also warn us when danger is present. A dog might bark at demons we can't see, perhaps one appearing as an animal. Once Lama Lhanang's mother saw a dog go into the neighbor's house and run out with a human heart in his mouth. No one else saw it, and the next day the neighbor died.

Lama Lhanang shared that he was attacked by a demon in the form of a cat one night. While sleeping, he was startled awake by a huge, black cat sitting on his chest. He became angry and punched it with his fist. The cat ran out the window. In the morning, Lama realized that the window was not open and that he had broken his knuckles. An infection set in, and a physician told him that his hand might have to be amputated. Lama meditated until it healed. His friend did divination and told him he had been attacked by a demon that wanted him and other teachers killed so he could acquire their power. Perhaps this explained why Lama Lhanang said he was "allergic" to cats.

During art classes, he acted like a young boy, turning his cap to the side and impersonating Gollum from *The Lord of the Rings*. He giggled and made shadow animals, such as llamas, with his hands, drew thumbnail sketches, and told stories about Buddhist deities. My favorite story was about Talkmed and Maitreya, the next Buddha to come.[47] A similar version of the story is found in *The Tibetan Book of Living and Dying*.

As Lama Lhanang told it, a man named Talkmed once wanted to see the Buddha Maitreya. So he went to the mountains to meditate for six years. In all that time, he never saw Maitreya, so he gave up and came down from the mountain. As he walked along the road, he saw a man polishing a metal rod. Talkmed asked what he was doing. The man explained that he had no needle to sew with, and so he planned to polish the rod until it was small enough to be a needle. Impressed by the man's patience, Talkmed returned to the mountains to meditate for three more years. Still, after that time, he did not see Maitreya and gave up. As he walked along the road, he saw a man hauling dirt. The man said that the mountain made too much shade on his house, so he was moving the mountain one bag of dirt at a time. Again, impressed by the persistence and patience of the man, Talkmed returned to the mountain to meditate for three more years.

However, once again, he did not see Maitreya, so he gave up and walked out of the mountains. Along the road, a dog barked at him. The front of the dog was normal but the rear end of the dog was dead and rotten with stinking pus and maggots. Talkmed was filled with compassion at the sight of it and wanted to clean out the wound and remove the maggots. Of course, he could not kill the maggots, so he decided to suck them out of the wound and spit them out on the ground. He held the dog and bent down to start sucking, when the dog turned into Maitreya. Amazed, the man said, "All this time I have wanted to see you, and now you come. Why did you take so long?" Maitreya said, "I have always been with you."

Lama Lhanang explained that because Talkmed had a pure mind, he was able to see Maitreya. Compassion allows us purity and the ability to see what is true. Perhaps my compassion for injured animals helps me contact their spiritual nature. Lama's story reminded me of an experience I had with a horse, so I told him about it and asked for his insight.

The Buddha-Nature of a Mare

It was late one Friday afternoon, about "beer-thirty," when Durango residents either start drinking or finally go home to do weekend chores. That was when Fred King discovered something he had been too busy to notice before and called me.

"Kiddo, I have a really bad wound that I need you to come treat right away," was all Fred said.

I drove into Fred's yard to find several solemn people standing around a sorrel horse. Fred, his wife, his daughter, and the man who owned the horse were silent. No one said a word. I could see from a distance that the mare's left forelimb was swollen to the size of a telephone pole. I walked around to inspect it and observed the leg to be a swarming cauldron of maggots. A laceration ran across the inside of the axilla (the "armpit" where the forelimb meets the chest) that left a wound as large as a football. Flies hovered over their offspring, the larvae ranging in size from barely visible threads to the size of puffed rice. Deep holes and crevices in the flesh teemed with little white larvae.

"Oh, maggots!" I said with glee. I used to be horrified by maggots, but after decades of practice, I had come to realize how helpful they are. They had saved this mare's life; she would have died from infection without their help. "Do you have any hydrogen peroxide?" I asked.

When I said this, Lama Lhanang's face winced in horror. I assumed he was concerned about the maggots, and I said, "Hydrogen peroxide won't hurt the maggots. It only bubbles them out of all their hiding places, and they fall on the ground. I just thank them, and the birds eat them later."

Fred's wife retrieved a couple of bottles, and so did I. "Okay, everybody, the family reunion is over," I said to the maggots. "Thank you very much for your assistance, your job is done here." I squirted the antiseptic into the wound using a 35-cc syringe, and then I scooped them out with a spoon.

A teaspoon is often the best tool to use for scraping maggots out of a wound. The larvae stick tight in between the skin and the muscle, which I found as I pulled the hide back with forceps, exposing a wiggling mass. I scooped the maggots out just like cleaning rice off the sides of a cooking pot.

Understand that this horse was not tied to anything, and I did not sedate her. She stood absolutely still as I irrigated the wound with hydrogen peroxide over and over again, and then dug out spoonful after spoonful from between tissue planes and picked out stubborn little fellows from tiny holes with forceps. It took at least an hour to get all the maggots out, and that mare never flinched. I cut dead tissues, muscle, and skin away, and she never moved. I washed the wound with soap and rinsed it with the hose. Her leg steamed, it was so hot with inflammation.

"Let's get her some grain and put this antibiotic powder in it," I finally said when all the cleaning was done. When we set the feed in front of her, she made her first move and started eating.

"She'll never be worth a damn, will she, Doc?" asked the owner.

"She'll be fine. Just feed this antibiotic and hose out the wound every day. It might take a month for the wound to completely heal, but after that, she should be good to go."

And she lived happily ever after.

The reason I tell this story is because Fred was amazed by the horse's behavior. Even though I hurt the horse, she was not afraid and did not resist. I see animals cooperate every day, and I have had similar experiences with maggots before, where a normally fractious horse has allowed me to help it.

Buddhism teaches that animals function out of passion, aggression, and ignorance. However, they do not always act like dumb animals. There are times when the animal nature ceases and the divine nature shows itself, when they act aware that I am helping them, when they show gratitude and compassion to other beings. I see this kind of "knowing" every day. I am blessed to connect with the Buddha-nature in animals because of my compassion for them combined with the need to be fully present with their suffering. Perhaps this is why those of us who recognize this quality in animals feel like the beasts are enlightened. We recognize their inner Buddha, just as Talkmed saw Maitreya.

I asked Lama Lhanang: "When I feel compassion rather than aggression, the animals become more cooperative. Does my compassion enable me to see Buddha-nature in the animals?"

Lama Lhanang said, "Yes, if you have tremendous compassion, you can talk to the animals and they can understand you. They can talk to you, too. And yes, everyone has Buddha-nature, even insects; they just have different body types. We are all the same consciousness or same soul. That is why we should never kill anything."

One night, at the retreat, we had a discussion with Lama Lhanang about mercy killing. He told us that some Buddhists could kill an animal and then bring it back to life without incurring bad karma. One man killed a deer and brought it back to life; another cut a cat's head off and then put it back on. However, he insisted that in order to be ordained as a Buddhist, one cannot kill. Jackson, a young man on the retreat, questioned Lama about this repeatedly. He was born into a family of hunters and fishermen. Those activities were important in his family life.

"What if you hit a deer on the road, and it is all mangled and dying. Can you finish it off?" asked Jackson.

Lama said, "I would not kill it. I would drag it off the road and say prayers."

Ariel asked whether it was okay to put dying pets to sleep "if it feels like the most compassionate thing to do."

Lama replied, "I would always choose not to kill, even to prevent suffering of a dying animal. I cannot choose for you; it is not my decision. Each person has to choose. I do not want the responsibility of saying what you should do."

As I've discussed, the issue of mercy killing arises constantly for veterinarians. About this, the Venerable Khenpo Karthar Rinpoche commented: "Ordinarily there is no such thing as mercy killing. Such things are done out of ignorance or what might be called 'idiot compassion.' Our knowledge is completely limited in this situation. We see only the temporary suffering the person might be going through. We have no understanding of the workings of karma, or what karmic conditions — possibly the most intense suffering — are waiting for this person after death."[48] Regarding animals, I wonder if an animal's karma may be less of an issue than a human's.

Sometimes even ordained Buddhists have killed when they believed it was for a higher good. For example, when a monastery in Woodstock, New York, was going to be condemned because it was overrun with cockroaches, the monks prayed for the roaches and had them exterminated.[49] This is analogous to killing *Streptococcus* bacteria to save a horse or a person who is dying from an infection. To never kill is impossible. Soap, antibiotics, windshields, automobiles, footsteps, pesticides, and dewormers all kill under our command. People have to reach into their own hearts to find the wisest, most reverent, and most caring action. I believe we must approach the killing we do without hatred, but with compassion. Whether we accept the Buddhist concept of karma and reincarnation or not, we need to reexamine the modern custom of euthanasia.

The Right to a Natural Death

Salty Bones was an emaciated, old horse that caused half the people in La Plata County to worry. People had telephoned the Humane Society for the past eight years complaining about this "neglected, abused horse." It was August 2003, and Salty Bones was thirty-six and a half years old, when the animal control officer called me.

"We need you to go tell Craig Walton how to get some weight on his horse, Salty."

"That horse is thirty-six and a half years old!" I replied. "He has no teeth and a heart murmur. You're not going to get that horse to gain weight."

"Have you body scored him?"

"Yes, he's a zero, a rack of bones, but we're not going to change that. How many horses do you know that are that old?"

"Well, Mr. Walton needs to feed him some senior feed."

"He does feed him senior feed. That horse is strong; he's hard to catch and even harder to examine. Craig's in worse shape than the horse is. All we're going to do by forcing feed on that horse is kill him with colic, maybe get Craig injured, and make him spend money he doesn't have."

"Well, if he doesn't get some weight on that horse, we are going to enforce euthanasia."

I got sassy. "So, are you going to start killing off old, skinny people, too? And how about charging people with cruelty and neglect for letting their horses get hog fat on grass until they can't walk because of laminitis? Those horses suffer a lot more than Salty is."

The officer said I cared more about the owner than the horse and decided to find another veterinarian.

The problem was not so much that Salty Bones was skinny, but that he lived along Highway 55 South. Every morning Salty and his two fat pasture mates stood next to the highway to catch the first rays of morning sun just in time for everyone to drive to work. In the previous year, complaints to animal control had come in weekly, sometimes daily, and the officers must have been tired of hearing about him.

Craig Walton was a retired IRS agent, an intelligent, independent, and somewhat ornery man. He was poor, largely due to his sickly wife's constant medical bills. They lived in a rundown trailer home, and although they owned a beautiful piece of property on the Animas River, trash from old rusty machinery and collapsed buildings littered the landscape. When I visited, I learned to stay on established tire tracks to avoid a flat tire. Amazingly, in the ten years I had been treating Craig's horses, I had only seen one wound, and that was not on Salty.

Craig had owned Salty Bones for the horse's entire life. I figured Craig must be doing something right for Salty to live so long. Until recently, Salty had belonged to Craig's son, who was born with a congenital heart disease. Salty had made it possible for the boy to ride in the mountains, but Craig's son had died that May. Craig thought the horse "deserved to live out his days," and I guessed the horse would not be far behind, but I was damned if the police were going to force the date. I was prepared to fight.

Craig was less defensive. He knew his rights and joked, "I'm going to hire a preacher, and put an ad in the paper inviting the county to see the euthanasia. Do you think that will make everybody happy?"

The animal control office called another veterinarian, Dr. Sanders, who called me to get the inside story. She, too, had a client with an old horse like Salty, only he kept a blanket on the horse so people wouldn't call the police. Dr. Sanders examined Salty with the animal control officer and determined that he had a heart murmur and no teeth and yet was strong and ornery and not in pain. The feed was adequate: lush pasture and senior feed daily. Dr. Sanders suggested a blanket, but Craig wouldn't go for it. No one was going to tell him what to do. It was August, and Craig did not think a blanket was needed, and he wouldn't agree to use one in winter, either. I never bothered to argue with him; he had enough problems.

One sunny day the following February, in 2004, Salty lay down in the dry grass and would not get up. Of course, he did it right next to the highway. He refused to eat or drink or even to lift his head. He was done. I got six phone calls from complete strangers wanting me to come out. I finally insisted that I had to hear from Craig. A woman said that he was right there, and he wanted me to come. Salty had finally decided to die. Now every passerby wanted me to hurry up and kill him before he died on his own.

It seems we cannot let an animal die naturally anymore. Not only is it considered inhumane to let an old animal die, now it is considered illegal. Unless an old horse drowns, like some have, or dies of colic in the middle of the night when no one is watching, we feel a desperate need to rush in and end life for them. Heaven forbid that an old horse should suffer a minute longer. Still, we allow all sorts of suffering to occur during life. It is only at the moment of death that suffering is abhorrent.

Let us face it, it is not the animal's suffering that we want to avoid. It is our own. We cannot stand to see an animal die. Death terrifies us, so we move it along gracefully and bury it quickly so we don't have to watch the natural process.

Buddhism offers another way. Be fully present with the suffering animal, thereby showing more compassion, transforming their pain, and doing more good. I do not expect the majority of people to believe this, but I hope that if someone else chooses this path out of love for their animal, more people will understand and let them be.

Craig was not Buddhist, but he was in no hurry to kill his old friend. I arrived at the Waltons' to find one woman on the side of the road fixing a flat tire on her truck. The old horse's heart sounded like a washing machine. Salty was unresponsive. When I asked Craig if he wanted me to finish the horse's life with euthanasia solution, two other strange women there nodded their heads vigorously. I waited for Craig to decide.

He told me to go ahead, and I gave Salty Bones a lethal injection. I thanked him for being such a good friend and told him to "go to the light." That was my custom long before I knew about the Buddhist belief that at the moment of death it is important to go into the clear light. I hoped the white horses that had carried Lonna away during her near-death would sweep Salty along with them into the "great white light."

In the end, the suffering of every person in that part of the county and in the sheriff's department was temporarily appeased. I did my job, and it may be that I have some karma to pay for all my killing.

If my karma means I return as some passionate, aggressive carnivore, I should like to come back as a house cat, perhaps in the home of an American Buddhist. Then, I can bring her gifts of live mice, snakes, and lizards, like my cats do. The Buddhist can reflect on dualism as she sets the critters free and spoils me with my every demand.

I met a cat like this when Lama Lhanang, Rachael Thompson, and I visited Tsultrim Allione at her home for tea. I sat in a soft upholstered chair with an end table to the left. A large, handsome, caramel-and-brown tabby cat climbed on the table asking to be petted. He had an unusual tail like a husky's that curled over his back. I know how to pet a cat, and as I scratched his ears and neck, he closed his eyes and pressed his head into my fingers, purring as he enjoyed the affection.

"Oh, he really likes you," said Tsultrim.

"He's a lover."

"He's my father," she said.

"Pardon me?" I wanted to know more, and Tsultrim went on to explain how she asked a Tibetan lama who reads incarnations where her father went after he died. The lama only spoke Tibetan but conveyed that her father was an animal with stripes on his face. Later the lama visited Tsultrim's home, and when he saw her cat, he said that it was her father. He lived a charmed life, if you ask me, one I would be honored to have.

Not all Buddhists believe in the possibility of reincarnation in an animal form. Zen Buddhism provides another idea about rebirth.

Zen Buddhism

An old pond, ah!
A frog jumps in:
Plop![50]

This haiku poem by seventeenth-century Zen poet Matsuo Basho is sometimes said to sum up the entire meaning of Buddhism. The old pond is emptiness. The frog represents conventional reality. The sound of water is the combination of the ultimate and the relative, the moment of awakening in the still pond of consciousness.

Poetry, paintings of landscapes, flower arranging, calligraphy, archery, and the martial arts are infused with the spirit of Zen, which emphasizes naturalness, spontaneity, and manifesting the Buddha-nature in daily activities.

Zen came about when Buddhism spread from India to China and then to Japan around the sixth century. The Ch'an school of Chinese Buddhism evolved into Japanese Zen. *Ch'an* means "meditation," and Ch'an tradition distrusted words, loved paradox, and emphasized the direct person-to-person transmission of insight.[51]

Nature became more significant in Zen as Buddhism evolved. In China, Buddhism mixed with the indigenous religion of Taoism, the two being similar philosophically. As it is said, "The Tao is empty like a bowl."[52] Taoism made nature a more important concept in Chinese Buddhism. Likewise, in Japan, Buddhism combined with Shinto, the indigenous shamanism, which worshipped *kami* or nature spirits. Along with this evolution came divergence.

Although Zen is part of the Mahayana sect of Buddhism, it is quite different from Tibetan Buddhism. Zen is not concerned with deep philosophical understanding. The emphasis is on sitting in meditation. One can never understand emptiness through mental discourse; it must be experienced directly, so the Zen master emphasizes practice, which consists of sitting in meditation and concentrating on your breath in order to achieve an awareness of emptiness. Zen practice itself is a form of enlightenment, since we each access Buddha-nature in practice. Enlightenment is awareness of one's true nature and experiencing oneness with all things, being free of the delusion of a separate self. The Zen definition of nirvana is simply being awake and aware in our daily life. Samsara is nirvana.

The realization of emptiness may also occur during other activities in daily life. For example, I experienced "emptiness" in karate training. My sensei taught me about a "no-mind" experience that one must have to be most effective in self-defense. We studied the teachings of Master Gichin Funakoshi (known as the father of karate) from Okinawa, who described karate as the empty-hand way. *Kara* means "empty." Master Funakoshi taught that "form is emptiness and emptiness is form; the form of the universe is emptiness."[53] According to Master Funakoshi, the four meanings of kara are as follows: 1) Karate is the empty-handed way. 2) Just as the clear mirror reflects without distortion, so the one who studies karate-do must purge themselves of selfish and evil thoughts. 3) Once you decide to stand up for justice, you must be hollow — kara — inside and straight, that is, unselfish and gentle. 4) The form of the universe is emptiness — kara. Thus, emptiness is form itself.[54]

While sparring with a woman, a black belt, I had my first insight into "no-mind." Then, I thought, "I'm doing it." At that moment, because I was thinking again, the woman punched me in the face. That taught me not to spar and think at the same time. Karate became a moving meditation for me. As painful as training was in the state of mindfulness, I do not remember feeling anything in no-mind.

The mind creates metaphysical questions. So Zen avoids those and focuses on what the Buddha said — focus on the moment right now. Therefore, Zen is indifferent to iconolatry and spiritual authority; it has no creed other than that the kingdom of heaven is in the heart of a person. It has no sacred texts other than to say, "The universe is the scripture of Zen."[55]

The Buddhist belief in the lack of "consciousness" in animals is seen in Zen as something of a good example. The Zen Buddhist knows that in order to recover the state of Buddhahood, to quote Coomaraswamy, "He must become again as a child, he must possess the heart of the wild deer; not withstanding he must also overcome the ignorance of which they are not yet aware." And also: "For in animals and children the inner and outer life are at one, the duality of flesh and spirit which afflicts us with a sense of sin are not yet felt."[56]

Zen Buddhist teacher Jakusho Kwong talks about a little finch that came to his window and sang brilliantly with no effort. He thought it sang so beautifully because it "did not have consciousness, so there was no self-concern."[57] He wished that people could experience that simple and pure thing, to become themselves, to sing like birds.

Natalie Goldberg, author and Zen practitioner, described the animal's consciousness to me by saying that they "do not accumulate thoughts." They let go of their thoughts, staying free of fabricated suffering, and live purely in the moment, expressing their true nature. The animals show us the ideal. They are true to who they are.

Thich Nhat Hanh agrees: "The historical Buddha is a beam of light sent into the world to help relieve suffering of living beings. The living Buddha is still available to us.... If that ray is not apparent to you, don't worry. There are many other rays, or transformation bodies, expounding the Dharma — the trees, the birds, the violet bamboo, and the yellow chrysanthemum."[58]

Zen koans are stories or questions that intend to make the student stop thinking and jolt them into emptiness. The classic example is the question: If a tree falls in the woods, and no one is there to hear it, does it make a sound? Shunryu Suzuki-roshi uses a frog koan as an example of how to be enlightened in zazen meditation: "The frog sits like us but has no idea of zazen. He does not think he is doing anything special, whereas we think, 'I am practicing zazen.' There is no Buddhahood beyond your ordinary mind. You should sit like a frog always. That is true Zen.... When a frog becomes a frog, Zen becomes Zen. When you are you, you have the practice of a frog. He is a good example of our practice. When you understand a frog through and through, you attain enlightenment. You are Buddha.... Even if you are sitting in meditation, your mind may be wandering. If we are like a frog, we are always ourselves."[59]

Zen Buddhists use the term *rebirth* rather than *reincarnation*, and they do not discuss rebirth in a physical way. Instead, rebirth is a union with all things.[60] A nine-year-old girl once asked Thich Nhat Hanh what he had decided to be in his next life. He was taken by surprise and said, "Well, it seems that I am paying so much attention to the present moment that I have not decided yet. But since you ask, I will try to answer. I will be many things. I will be a cloud; I will be a bird. I will be a butterfly, and look, I will be a little yellow flower like this one in front of us."[61]

The Zen view of killing animals also derives from the Buddha's teachings on the Eightfold Path, which includes right action: not killing, not stealing, not misusing sex, not telling lies, not getting angry, and so on. For many Zen practitioners, not killing means not taking life unnecessarily.[62] In our daily lives, we kill unintentionally, such as bugs under our feet or in our vegetables as we clean them, and we kill intentionally. The important thing is to be mindful of our actions and compassionate at heart.

Buddhism has helped me be more joyful in my life. I know that I can worry, hate, blame, and dwell on ugly images, creating more unhappiness. Or I can take the opportunity to face reality just as it is, with all of its bloody pieces of meat, to look deeply at it with loving compassion, and then rise like a swan off the lake, not looking back and not suffering.

Not only does Buddhism tell us that animals feel compassion, but according to the Jataka tales, animals teach humans about compassion as well. Animals have the same inner light as humans, although they are not caught up thinking about it, not conscious of it. They are themselves, perfect bodhisattvas.

My direct experiences as a veterinarian have enlightened me to the presence of the Buddha-nature in everything. In practice, I am forced to be fully present with gore and death. When I let go of opinions about good and bad, I connect with the animas of the animals and their humans; my anima meets theirs, and I realize the bliss called equanimity. I may not always understand what I read in *The Tibetan Book of Living and Dying* about aversion, compassion, and equanimity, but I do understand when a stinky, old dog that can barely walk comes into my house with a large tumor on his side, oozing bloody serum all about the place, and the people cannot bear the idea of euthanasia, and they want my help. I must get past aversion and repulsion and accept my duty. When I choose to soothe rather than judge the situation, I move into acceptance; from acceptance, I find myself caring, which is compassion. From caring, I connect with the Buddha-nature and feel the joy of equanimity. Then I heal in my work of healing others.

Any merit gained from these practices is dedicated to the benefit of all sentient beings.

CHAPTER 7

Islam, Judaism, and Christianity: The God of Abraham

In the excerpt from the ancient Hebrew literature above, we see how God describes himself to Moses. These words provide more wonder than definition. Words simply fail to define God. Even so, about 93 percent of Americans say they believe in God or a Universal Spirit.[2] About 55 percent of the people on earth (Muslims, Jews, and Christians) worship this God, who is historically referred to as "the God of Abraham."[3]

The histories of Islam, Judaism, and Christianity all begin with Abraham (Ibrāhīm in Islam). He was both the patriarch of the Jewish people and the original Muslim.[4] His two sons, Isaac and Ishmael, became the forefathers of the Jewish and Muslim religions, respectively. Christianity developed out of Judaism, since Jesus Christ was Hebrew. The holy books of all three religions (Qur'an, Tanakh, and Christian Old Testament) refer to their God as "the God of Abraham." All three religions tell the stories of Adam, Abraham, Moses, and Jesus.

Believers of these three religions agree that God is the creator of the universe. However, beyond the definition of *creator*, opinions differ about the nature of God, not only among the three religions but also within their numerous sects and denominations. Most view God as omnipotent, omnipresent, and omniscient. The beliefs of these monotheistic monuments stand like thick, snarled hedges. My mission to explain what these three religious groups say about animals is like trying to untangle a strand of fur from a hedgerow. The complexity of this makes it both challenging and fascinating. As elsewhere, my intention is to present each religion objectively, on its own terms. However, in this chapter, I feel obliged to offer my own opinions because I know this God.

My Father Who Art in Wisconsin

I was born to the son of an Evangelical preacher in a small town in Wisconsin. My father held my infant body over the baptismal font as Pastor Patterson splashed water on my head, baptizing me into the Trinity Lutheran Church, in the name of the Father, the Son, and the Holy Ghost. My mother, Gloria, prayed for me in German, asking God to send a guardian angel to protect me "from all harm." Growing up, we prayed all the time, at every meal and at bedtime.

My first Bible, the Children's Living Bible, had Jesus on the cover carrying a lamb. Inside, colored pictures showed Noah leading animals into the ark; ravens bringing food to Elijah in the desert; Jonah spit out by the whale; and Daniel safe among the lions in their den. It looked to me like God and his animals were close friends who worked together.

Every Sunday we went to church, first to Sunday school and then to church service, where Mom sang. She stood barefoot in the balcony belting out the Lord's Prayer in a vibrato that rattled the stained-glass windows, reaching a high C note that woke the elderly Norwegian farmers snoozing in the pews. She also sang at weddings and funerals. One of her favorite songs was "His Eye Is on the Sparrow," which is from a Bible verse saying that God knows about every sparrow that falls, so we know he watches us, too.

In the summer I went to vacation Bible school, and as I got older, to Bible camp. In the summer of my ninth year, I spent a weekend at my grandmother's cottage on Lake Winnebago, outside of Oshkosh, Wisconsin. In the evening, my cousin and I caught lightning bugs and kept them in a jar. I set the jar on the table next to the hide-a-bed and watched the beetles blink as I said my bedtime prayers. That night, I also prayed out loud for my dog, Sparkle, to go to heaven when she died. Imagine my surprise when Grandma said, "Acht, dogs don't go to heaven. Animals don't have souls." I said nothing; I knew she was wrong.

The fireflies glowed like the halo over Jesus...and they didn't have souls?

God had Noah build a huge ark to save the animals from the flood, and after they were back on land, God made an agreement (a covenant) with the animals that he would never destroy them all with water again. He created the rainbow as a sign of this covenant. Yet he wouldn't save them after they died? His eye is on the sparrow, but he would abandon it at death? Based on biblical stories, it seemed to me that humans and animals are literally in the same ark. I could not believe that animals did not have souls and that they did not go to heaven. The Bible makes it very clear that God loves his animals.

In the Wisconsin summers I caught every kind of small creature I could — snakes, turtles, frogs, and insects. One night, during my parents' Bible study in our living room, I heard Mom screaming. I was upstairs in bed when Dad yelled, "Karlene, get down here right now! Your snake is loose." The women stood on the sofas as I crawled around the floor in my nightie looking for the poor creature, which terrified them. You would have thought it was the devil in the Garden of Eden, and that I was an "evil" Eve. For heaven's sake, it was just a little grass snake, and I was just a little girl. Adults seemed to always blow things out of proportion.

My father found it challenging to keep me out of the swamps and other dens of iniquity. He was the president of the church, and in my teenage years, he taught my Sunday school class, one of the most embarrassing things I could imagine. That year I became an adult member of the church at confirmation. I took my first communion, drinking the wine that is "the blood" and eating the wafer that is "the body" of Christ. Then, Dad enlisted me to teach Sunday school as well. He never made my brothers be so involved in church; I rarely did anything else. Each night after dinner, I had to read the Bible. My friends were outside playing "capture the flag" while I was reading aloud Kretzmann's *Popular Commentary of the Bible*. Could I go to the youth center on Thursday nights? No! My father insisted that I go to Luther League instead. How boring! Crabby junior pastors and prune-faced, old women complained about the "evil" things my friends were doing at the dance halls, where I wanted to be — listening to music, having fun, and laughing. My friends were good kids. So to have some self-righteous "Christians" condemn them really turned me off.

It was the sixties, and my mother, an interior designer by profession, wore the first pantsuit to church. Our conservative congregation expected women to sit still in a proper Sunday dress on hard, wooden pews. My mother seemed inappropriate marching around in her purple Nehru jacket and matching bell-bottoms. The Lutherans looked down their collective nose, but Gloria kicked off her flowered sandals and sang as usual. They finally got used to her and even followed her fashion trends.

I will never forget the night my mother answered the telephone and cried out, "Who is this, who is this?" Someone called to threaten the family furniture business. There had been a political disagreement with some of the members of the church. My father sided with Pastor Patterson, a gentle, kind man who always smiled, and he was a horseman, too. He raised Shetland ponies, and my dad bought one of them for me. So, my father appeared right in my eyes; he abided by the church bylaws in this circumstance, which made some members of the congregation angry enough to threaten Gloria, anonymously. They said that our family store would fail. Then these unnamed individuals left the congregation and started their own church. Jesus was all right with me; Christians, however, were not much like Christ. My father dismissed the problem, saying, "Church is for sinners, not for saints."

To me, this is the perfect example of what happened in the history of the Christian church: Endless disagreements created varieties of churches with differing beliefs, all called "Christian." Christianity is the largest religion in the world, with Islam a close second. I know many good, loving people who practice what they preach by giving generously to the poor, loving their neighbors, and even loving their enemies as Jesus taught. On the other hand, too many who call themselves Christians are full of hate — Protestants and Catholics hating each other; Christians hating Jews, even though Jesus, his mother Mary, and the man who wrote most of the books of the Christian Bible, St. Paul, were all Jews. There's too much murder in the name of God and too much self-righteous condemnation of others. As an adult, this hypocrisy drove me away from church.

I went to the mountains to find God in nature, with the animals. Today, I still talk with God daily, mostly saying, "Thank you!" I feel the divine presence everywhere, although my understanding of God may be vastly different from that of some Christians. Most importantly, after learning about the problems of dualistic thinking, I realize that I too must not judge. I have visited many kinds of churches, and I resonate with all those that preach love, including native, pagan, and Hindu teachings. The Agape International Spiritual Center and the Unitarian Universalist Fellowship are two examples of love-based religious groups I enjoy that accept a wide group of religions.

With this in mind, I will look at what is said about animals in the Bible (using the Revised Standard Version, unless otherwise stated). The Bible contains many wonderful stories about animals with important lessons. First, though, it helps to understand a bit about how the Bible was written and its interpretation today.

The Bible and Biblical Interpretation

According to Phillip Cary of Yale Divinity School, the books of the Bible were written in ancient foreign languages that we can no longer translate with complete accuracy. We do not have the full linguistic or cultural code that would enable us to be as supple and sophisticated in our understanding as we would like. The Bible is a compilation of ancient stories. No original manuscripts of the Bible remain. The earliest writings found of the New Testament are tiny scraps the size of credit cards. For the most part, biblical writings are copies of copies of copies of the originals, which were handwritten by scribes who, like all human beings, made mistakes in spelling, omission, and translation. They also changed things of their own accord. Of these copies, we have 5,400 different Greek manuscripts of the New Testament, not counting the thousands of versions written in Latin, Syriac, Coptic, and Armenian.[5] From these, numerous translations and interpretations of the Bible have been rendered and exist today. In 1707, the Oxford scholar John Mill compared a hundred of these manuscripts and found thirty thousand "significant" places of variation among the translations.[6]

Given this, how does one take the Bible literally to be the word of God? Fundamentalist Christians believe that the Bible is the divinely inspired word of God, and for them, it does not matter what scholars or scientists say about the historicity of the book. I tend to agree. The book each believer holds in his or her hands, and the teachings each person receives, is the important thing. I, however, find the history quite interesting.

What surprised me most in my study of the Judeo-Christian religions is how much of what we think of as the biblical God's truth is really Greek philosophy. Greek philosophers, such as Plato, theorized about the concept of the soul; they imagined the soul being separate from the body at death and going to heaven. Ancient Hebrews did not believe in life after death; some modern Jews still do not. In the Old Testament, Isaiah 57:9 mentions Sheol, a place under the earth, neither heaven nor hell, where the dead go to rest.[7] The idea that souls go to heaven after death came later in history from Plato and other Greek philosophers, as did the words *devil* and *hell*.[8] Scholar Bart Ehrman agrees, explaining that the philosophy about heaven and hell transpired later in Christianity, when the apocalyptic vision of the earliest Christians failed to materialize.[9] Phillip Cary wrote, "The notion that good people's souls go to heaven when they die is not found anywhere in the Bible."[10]

Reverend Ed Beck, a Methodist, explained to me that in the Bible only two people ascend to heaven, Jesus and the prophet Elijah. Both go bodily; the body and soul do not separate. Even in the Qur'an, Jesus does not die but goes directly to heaven.[11] In the New Testament, there is no body of Jesus in his tomb. He is "risen," body and soul. When Jesus appears to people after being raised from the dead, he is in a body; he is not a spirit. See Luke 24:36–40. According to Reverend Beck, Christianity does not believe in the eternality of an immortal soul, but rather in the resurrection of the body and everlasting life, which sounds to me like reincarnation. Reverend Beck wrote to me, "As St. Paul would say, 'You are raised in a new body.'"[12]

Greek culture was the dominant, intellectual force at the time the New Testament books were written, and the Greek philosophers were influential. Greek-speaking Jewish philosophers found the modern thoughts of their time to be the pinnacle of wisdom, and they incorporated the ideas of Plato and Aristotle into their biblical interpretations. This has continued over the centuries, and thoughts about God evolve. When reading Bible stories, one has to remember the perspective of the writers. Let us now consider what the Bible says about the spiritual nature of animals.

Animals in the Old Testament

Paganism is evident in the first book of the Bible — Genesis. One may accept these stories from the tribal and pagan perspective in which they originated, understanding that blood sacrifice and multiple gods seemed normal. Christians, however, tend to interpret the ancient Hebrew scriptures while imposing modern Christian beliefs over them. Orthodox Jews read the stories differently. The sacred myths found in Genesis compare with other Middle Eastern stories of the time, such as the Gilgamesh epic. Various cultures in the Mediterranean had the same creation images, a flood story, and similar heroes and mythical tales with animals speaking. The creation legend of Adam and Eve in the Garden of

Eden is an example of a folktale about the beginning of time that reflects a tribal and agricultural people.

Genesis presents two different creation myths. Genesis 1 perhaps shows the polytheistic influence, with God speaking in plural: "Let us make man in our own image." Rabbi Gershon Winkler provides an interesting reason for the use of the word *us*. In this version, God creates the animals first. Winkler quotes the ancient rabbis: "When it came time to create the human, the Creator addressed *all* that had been made until that moment, both in the sky and in the earth, and said to them: 'This final creature is too complex for any of you to bring forth alone. Therefore, let us all join together in its creation. All of you join in making its body and I will join you in making its spirit.'" This implies that the creator addressed all of creation before making the first human. God incorporated all of the attributes of all the animals and plants and minerals, and so on, that were created up to this point. "In each of us, then, are the attributes and powers of all the creatures of the earth."[13] Thus follows the belief that any animal one sees represents a reflection.

After creation, this human — whose sex is not named, since there is no duality yet — is given dominion over the fish, birds, and every living thing. Orthodox Judaism envisions a hierarchy in creation, with humans being expected to exercise responsibility for the animals as stewards, always with kindness.[14]

In Genesis 2, God creates man first, then the animals. He makes both from the earth. The Hebrew word for *Adam* means "ground" or "arable soil."[15] Adam was made "from the dust of the ground." Humans are *adamah* — "earthlings," made as part of the world body. God also makes the animals from earth. See Genesis 2:19. Then God makes a woman from Adam's rib. The Hebrew word for *Eve*, *Chaya*, means "life." According to the Hebrew scholar Amy-Jill Levine, Eve is the Mother Goddess of all life.[16] In this typical pagan agricultural tale, God put man into the garden to till it and keep it (Genesis 2:15). In this second creation myth, Adam names the animals.

Genesis 2:7: "Then the Lord God formed man of dust from the ground, and breathed into his nostrils the breath of life; and man became a living being." Genesis 2:18–19: "Then the Lord God said, 'It is not good that man should be alone; I will make him a helper fit for him.' So out of the ground the Lord God formed every beast of the field and every bird of the air, and brought them to the man to see what he would call them; and whatever the man called every living creature, that was its name."

Because the Hebrew words allow for a variety of meanings, Winkler reads the scripture differently: "Conversely, according to the Hebrew wording of the Hebrew creation story, it is just as feasible that it was rather the primeval *animals* who named the primeval *human*...because way back at the beginning of time, the primeval animals blessed and initiated the primeval human with each their particular nature and wisdom."[17]

The Great Flood and Noah's Ark

Genesis also contains the flood narrative. The story describes how people broke the covenant with God and were punished with a great flood. God tells Noah to build a giant ark to hold representatives of all the animals. Genesis 7:2–3: "Take with you seven pairs of all clean animals, the male and his mate; and a pair of the animals that are not clean, the male and his mate, and seven pairs of birds of the air also, male and female, to keep their kind alive upon the face of the earth."

No explanation of clean or unclean is given in the Bible prior to this; however, clean animals were used for sacrifice and could be eaten later on. There had to be more than two of the "clean" ones because once the water receded, Noah made a blood sacrifice of one of each of them to God. Genesis 8:21: "And when the Lord smelled the pleasing odor [of the burnt offering], the Lord said in his heart, 'I will never again curse the ground because of man.'"

From then on, God gave humanity permission to eat meat, with one restriction. Genesis 9:3–4: "Every moving thing that lives shall be food for you; and as I gave you green plants, I give you everything. Only you shall not eat flesh with its life, that is, its blood." This is part of kosher law and why, to prepare kosher meat, the blood is drained first. The ancient Chinese also drained and offered the blood in sacrifice. Likewise, the Qur'an describes survival from a flood and forbids the eating of blood; it gives humans permission to ride horses and to eat meat, that is, other than swine or meat that has been killed improperly. The meat must be pronounced in God's name, and one is to be grateful.[18]

Then God made a contract with the humans and the animals. Genesis 9:9–13: " 'I establish my covenant with you and your descendants after you, and with every living creature that is with you, the birds of the air, the cattle, and every beast of the earth with you, as many as came out of the ark. I establish my covenant with you, that never again shall all flesh be cut off by the waters of a flood, and never again shall there be a flood to destroy this earth.' And God said, 'This is the sign of the covenant which I make between me and you and every living creature that is with you, for all future generations. I set my bow [rainbow] in the cloud, and it shall be a sign of the covenant between me and the earth.'"

God valued the animals enough to save them along with Noah's family. He saved "unclean" animals not used for food, so food was not the reason he saved them. He preserved animals when he destroyed all the rest of humankind. Importantly, God made an eternal covenant with the animals as well. Clearly, animals are important to God.

God also gave the Jews laws regarding animals in the Torah, summarized by Rabbi Roller: "Under Jewish law, animals have some of the same rights as humans do." Animals must rest on the Shabbat the same as humans; we must allow an ox and human workers to eat from the produce they are harvesting. We are required to relieve an animal of its burden. We are not permitted to kill an animal in the same day as its young, and we must send away a mother bird when taking the eggs. "In fact," Rabbi Roller said, "the Torah specifically says that a person who sends away the mother bird will be rewarded with long life, precisely the same reward that is given for honoring mother and father. This should give some indication of the importance of this law."[19]

Winkler references Leviticus 17:3–4 regarding sacrificing animals, writing, "Killing an animal outside the complex rituals connected to the Sacred Altar was considered tantamount to murder."[20]

As mysterious as God is, he shows a great deal of concern for the beasts of this planet. I wonder why he would bother with these considerations if animals were just physical flesh and without spiritual content. Bible stories about the prophets show evidence that animals do indeed have a spiritual nature.

The Prophets and Balaam's Ass

The prophet is a conduit between God and the people. Biblical prophets are messengers, possessed by the Holy Spirit, who speak on behalf of God. Today, we use the term *channel* to describe a person who speaks for God or other nonphysical entities. Such activity is common in biblical literature, as it is in the Qur'an, which came as a direct transmission from Allah to the Prophet Muhammad. In the New Testament, people "speak in tongues" (glossolalia), transmitting information from the Holy Spirit, which still occurs today.

Prophets, seers, and oracles who went into trances and spoke ecstatic utterances were common in pagan times. Mediums performed divination through astrology, necromancy, casting lots, and hepatoscopy (examining the liver for signs). All these esoteric practices are common in the Hebrew scriptures.

Kabbalistic Jews define prophecy as a mystical state. The prophets often engage in soul travel to other realms, and that prophetic state is a sort of dying, as in shamanism. Rabbi David Cooper explains that the statement "man is made in God's own image" means that, at the level of human consciousness, humans have the potential for God consciousness or prophecy.[21] Rabbi Gershon Winkler calls the prophets "shamans."

The following story of the prophet Elisha's magic may inspire respect (2 Kings 2:23–25): Some boys were harassing the prophet Elisha because he was bald. "And when he saw them he cursed them in the name of the Lord. And two she-bears came out of the woods and tore up forty-two of the boys."

My favorite Bible story is about a prophet named Balaam and his donkey, often called "Balaam's Ass." Here, I paraphrase the story from Numbers 22:1–33:

> At that time, the tribes of Israel had fled slavery in Egypt and formed camps in the plains of Moab. There were thousands of them, and the people of Moab feared them, saying, "This horde will now lick up all that is around us, as an ox licks up the grass of the field." The elders of Moab decided to chase the Israelites off. They planned to do this by hiring the prophet Balaam to curse them.
>
> Balaam saddled his donkey and went with the officials of Moab. But God was angry because he went, and he placed an angel in the road as his adversary. Balaam was riding on the ass, and the ass saw the angel of the Lord standing in the road, with a drawn sword in hand, but Balaam did not see it. The ass turned aside out of the road and went into the field, and Balaam struck the ass. Balaam tried to drive the donkey down a narrow path between two vineyards, with a wall on either side. And when the ass saw the angel of the Lord, she pushed against the wall and pressed Balaam's foot against it, so he struck her again. Then Balaam rode the donkey in a narrow place, where there was no way to turn either to the right or to the left. But when the ass saw the angel of the Lord, she lay down under Balaam. Balaam's anger was kindled, and he struck the ass with his staff.

Then the Lord opened the mouth of the ass, and she said to Balaam, "Why have you struck me these three times?" And Balaam said to the ass, "Because you have made sport of me. I wish I had a sword in my hand, for then I would kill you." And the ass said to Balaam, "Am I not your ass, upon which you have ridden all your life long to this day? Was I ever accustomed to do so to you?" And he said, "No." Then the Lord opened the eyes of Balaam, and he saw the angel of the Lord standing in the way, with his drawn sword in his hand; Balaam bowed his head and fell on his face. And the angel of the Lord said to him, "Why have you struck your ass these three times? Behold, I have come forth to withstand you because your way is perverse before me. The ass saw me and turned aside before me these three times. If she had not turned aside from me, surely just now I would have slain you and let her live."

This story brings up two important points about animals. First, the donkey saw the angel of the Lord, and the prophet did not. How can a being without a spirit see a spiritual being? Indeed, shamans, Tibetan Buddhists, and many Christian and Hebrew teachers agree that animals can see the spirit realm. Rabbi Winkler's sources state: "The animals possess greater abilities to perceive the spiritual than do humans."[22] This is why animals have extra knowledge of impending phenomena such as tsunamis and earthquakes.

Scientific evidence also supports the idea that animals perceive energetic messages we humans seem unaware of. A research project at the Swedish University of Agricultural Sciences demonstrated that if a person leading or riding a horse has an increase in heart rate, the horse has an increase in heart rate.[23] Dog owners agree that a dog immediately knows if it likes someone or not. The lesson of the story is: The next time an animal you know well acts strangely about something, instead of getting angry, pay attention.

The second point is that the donkey spoke. A common theme of creation myths recurs; humans and animals could once communicate with one another, just as the snake spoke to Eve in the Garden of Eden.

In the *Journal of the Christian Veterinary Mission*, Dr. Kit Flowers reflects on the story of Balaam's ass. He asks us to think of times when the Lord has spoken to us through an animal or through an interaction with animals. He says that "animals are a part of creation where God's invisible attributes are seen."[24]

The ass spoke. Like a prophet or a charismatic, the Holy Spirit put words forth from the donkey's mouth. Surely, something that is only ignorant matter would not see angels and verbalize God's message. The animals communicate with God willingly; it is natural for them. They do not argue. As an example, see 1 Samuel 6:7–12 for the story of how the Philistines' cows willingly left their calves behind in order to carry the stolen Ark of the Covenant back to the Israelites.

In the spring of 2016, I discussed this point with a Benedictine monk at the Monastery of Christ in the Desert, near Abiquiú, New Mexico. The head padre invited Jean-Luc and me to meet the horses, and he told us on the way to the horse corral about Rusty, an old mustang used in several movies. Rusty was too smart; he could open any gate and was teaching the younger horses naughty tricks, like how to get into the gardens. At the corral, we met the wrangler monk, a burly man in appearance and personality. Instead of a long, black robe, he wore a short, black tunic with a thick, leather belt and jeans. The wrangler was in the process of hanging a new gate, and he introduced us to each horse and the dog that lived in his shed. He also shared stories about each of the animals he cared for, including the escapades of Rusty getting into the garden and other places by opening gates. When I mentioned, "Animals see the spirit realm," he replied, "Oh, I have no doubt about that! We have an angel guarding the monastery, and every time I take horses by that place, they always shy to the side."

I posed the question as to whether animals act willingly to do God's will or if they are puppets in God's hand. "Animals are more like puppets than we are," said the monk.

"Right, because we argue, and they don't!" I said.

We debated the concept of a hierarchy of spirit; the monk believed a human's spiritual nature was higher than an animal's. "Man has reason, *logos*; the animals do not," said the monk, as Jean-Luc helped him mount the gate to the post.

Then the padre said, "Rusty is watching you, Brother." We all laughed, knowing Rusty reasoned just fine. As we walked back to the chapel, I told the padre about the intelligence of the octopus: "They have no vertebrae, donut-shaped brains, and the ability to camouflage various parts of their bodies into different colors and patterns. An octopus blends into the surroundings and could be right next to you, and you wouldn't see it. They solve puzzles easily and squeeze through tiny openings, which makes it extremely difficult to keep one in captivity." Nodding, the padre smiled and said he had just read about Inky, an octopus that escaped from a New Zealand aquarium.

Humans learn more about the intelligence of animals every day, and we are discovering that their reasoning abilities are comparable to our own. Perhaps humans need to reconsider the idea of spiritual superiority.

Do Animals Have Souls?

*In looking into the eyes of an animal we may be aware that there is
something sacred and holy, something divine, in the animal. God's Spirit is in her,
shining through her, even as she is more than Spirit. It is as if she is a holy icon,
a stained-glass window, through which holy light shines.*
— JAY McDANIEL[25]

Christians who believe animals do not have souls claim that only Adam received the breath of life, "breathed into his nostrils" by God. This breath is the Holy Spirit, which supposedly the animals did not receive. However, animals clearly breathe and have life. So, if God did not give them the breath of life, where did it come from? Nowhere in the Bible does it say that animals do not have souls or eternal spirits. On the contrary, Ecclesiastes 3:19 states: "For the fate of the sons of men and the fate of the beasts is the same; as one dies, so dies the other. They all have the same breath, and man has no advantage over the beasts; for all is vanity."

Humans are so vane we think we are superior, when in fact we are equal in God's Spirit. Rabbi Gershon Winkler agrees, quoting ancient sources: "You are the Breath of God, of the Sacred Wellspring. The God Breath is the evolutionary spiral that emanates from God's will that there be creations and culminates in the manifestation of that breath/will as stone, plant, elk, lichen, sun, moon, and you. The tree or grasshopper that you pass as you take your daily stroll is therefore also the Breath of God. All that exists is being breathed into being or it wouldn't be." Winkler continues: "In other words, what you have been spending thousands of dollars in books and workshops to discover about your essential selfhood can be discovered gratis in the next animal you run into or that flies over your head. Both you and they carry the mystery of the other. Both you and they are the masks of the other. You are they in human clothing; they are you in animal clothing, or plant clothing, or rock clothing."[26]

Watchman Nee wrote that animals *do* have souls but *do not* have spirits. Nee said that only a human has "spirit," the breath of God; "God is Spirit."[27] This distinction rests on the translations of the Hebrew words for *spirit* and *soul*: *ru'ach*, usually translated as spirit, means "breath" or "life-force," and *nefesh*, translated as soul, means "vitality" or "that which breathes."[28] However, there is no dichotomy of body and soul in the original Old Testament. Still, various Jewish rabbis disagree on whether animals have souls or not. For instance, these quotes were posted on the (now-defunct) website AskARabbi.com:[29]

> RABBI DAN: My point of view is of one soul manifesting in every phys-ical body that inhabits this planet. As such, the soul is fragmentized but not individual. We do not "have" (a) soul.... It has us as its container through which to have a human (or animal) experience.

> RABBI FINMAN: Sorry, but since animals do not possess a higher soul, only a lower soul, they do not have an afterlife.

> RABBI BEN-MEIR: I believe that god only breathed His "Ruach" [*ru'ach*] into man but that gives me pause to think: is there a burning "nefesh" inside the creatures we have been given majesty over? I breed Shelties and I love these sweet, giving creatures. What do you think?

For me, the deciding verse is Ecclesiastes 3:18–19. I quote 3:19 above, and here it is with verse 18: "I said in my heart with regard to the sons of men that God is testing them to show them that they too are beasts. The fate of the sons of men and the fate of the beasts is the same.... They all have the same breath [*ru'ach*]." Numerous Bible commentaries on this verse continue to place humans above beasts, including the one by Nee. They seem to overlook the point made by the speaker — Ecclesiastes, the preacher. The Book of Ecclesiastes begins, "Theme: vanity of human wisdom." He clearly says that human vanity leads us to believe we are superior and that we are being tested to recognize that we, too, are animals.

Rabbi Gershon Winkler agrees. "Both you and they share the same life breath of Creator, and neither of you is any more important to Creator than the other."[30] The Qur'an (6:38) shares a similar conclusion: "There is not an animal on earth or a bird that flies on its wings which is not community with you... and they shall all be gathered to their Lord in the end."

In January 2008, two young Mormon "elders" came to my house. They wore black suits and ear muffs, but no coats. I scolded them for not dressing warmly enough and offered them a hot tea or coffee, which they refused, since a good Mormon does not partake in such stimulants. They showed me their version of the first book of the Mormon Bible, called "Moses," which describes God breathing life into the animals. Mormon founder Joseph Smith taught that animals have souls. Moses 3:19 states that God created all things spiritually; he created animals out of the ground and "breathed into them the breath of life."

While waiting for a flight in the Atlanta airport, I once asked an Episcopalian priest if he thought animals have souls. He said, "If you read the Book of Revelation, you will see that there are all kinds of strange creatures in heaven. If God can make room for them, I don't see why he can't make room for my dog." Another Episcopal priest told my mother that animals absolutely have souls and go to heaven, saying, "They will be right there along with us."

Some Catholics subscribe to the teachings of St. Francis of Assisi, who talked to the animals. St. Francis preached to the birds because, as it says in the Gospel of Mark, Jesus commanded the apostles to preach to all creatures.[31] St. Francis also spoke to a wolf that was harassing a community. In the story of the wolf of Gubbio (in Italy), Francis made a deal in writing with the wolf, who was killing people, in which the people agreed to feed the wolf in exchange for peace. Francis said, "Nature is a series of footprints that leads you to God."[32] For Francis, God's goodness was revealed in nature, so by contemplating a rock or a flower one could move from creation back to the creator. Just as St. Francis did, St. Basil also referred to animals as our brothers in his prayer: "Enlarge within us the sense of fellowship with all living things, our brothers the animals to whom you gave the earth as their home in common with us."[33] On December 5, 2014, Pope Francis quoted the apostle Paul saying, "One day we will see our animals again in eternity of Christ. Paradise is open to all God's creatures."[34] Francis reportedly wrote in the *Laudato si'* that "animals will join humans in the kingdom of heaven."[35]

The Anglican priest Andrew Linzey states, "There is no human embodiment totally unsimilar to the flesh of other sentient creatures."[36] We are literally in the same boat, the same bodies, with the creator's same breath. Everything that breathes has a soul, and everything alive has spirit, including the lightning bugs I caught as a child. Even individual cells breathe; they have organelles of respiration — mitochondria. I believe every living cell contains God's Spirit. It seems obvious that animals have a spiritual nature. It is plain for all with eyes to see, even crusty cowboys.

One autumn day I went to euthanize a horse named Goldy. He had Cushing's disease and was in his twenties. The owners had cared for him well until they went into bankruptcy. They sold their healthy horses and hired me to kill Goldy. A local, middle-aged cowboy, Stan Steed, dug a hole with a backhoe. It seemed like Stan didn't think much of me; he never addressed me directly. Instead, he acted as though I did not exist. I gave Goldy the fatal injection, and the horse fell down next to the hole. I listened to his chest with a stethoscope, and when the heart stopped, I pronounced him dead. Stan stood there staring at Goldy's body, and then he spoke:

"Now, I am not a religious man. But I take my son to Sunday school because he wants to go, and because I think it's a good education. But those Christians tell him that animals don't have souls. Now, you can't tell me that there wasn't something there a minute ago that isn't there now."

I agreed with him and tried to engage him in conversation about it, but he ignored me and got into the backhoe to finish his job. His point was clear, though. "Something," the breath of life, the life force that we call soul or spirit, was gone. I wondered where it went.

Do Animals Experience Life after Death?

Adam, Noah, and Jesus are the three heads of the three Earths. When Adam was
created, God surrounded him with animals. When Noah was delivered
from the Flood, God surrounded him with animals. When Jesus was born,
God surrounded him with animals. When Jesus establishes the renewed Earth,
with renewed men and women, don't you think he'll surround
himself with renewed animals?
— RANDY ALCORN[37]

The Bible provides good evidence for the notion that animals are reborn after death. For instance, consider Isaiah 65:17, 25: "For behold, I create new heavens and a new earth; the former things shall not be remembered or come to mind....

The wolf and lamb shall feed together, the lion shall eat straw like the ox." Here, God is speaking through Isaiah and describing this kingdom as a place where humans and animals all live together in peace and all are vegetarians, like the paradise times. Also consider Isaiah 11:6–7: "The wolf shall lie down with the lamb, the leopard shall lie down with the kid, the calf and the lion and the fatling together, and a little child shall lead them. The cow and the bear shall feed; their young shall lie down together; and the lion shall eat straw like the ox." Furthermore, Psalms 36:6 states, "Man and beast thou savest, O Lord."

The last book of the Bible, Revelation, is the vision of a prophet named John. In this apocalyptic image of the last days on this earth, Christ is a lamb sitting on a throne. Later, the Messiah rides a white horse (19:11). Every creature sings and gives praise to the lamb (5:13), and an angel calls to all the birds in mid-heaven (19:17).

Somehow animals appear in the new paradise. Either the animals are saved by God, as happened during the flood, or they are reborn into a new earth along with humans, which sounds like reincarnation into a new nefesh. One client told me that animals do not go to heaven, and although there are animals described in heaven in the Bible, those animals are not the same animals as those on earth with us now. This only goes to show that, as the scholar Bart Ehrman puts it, "Evidence can be read any way a person likes."[38]

Angels and Animals

Whether moving creatures or the angels: for none are arrogant (before the Lord).
They all revere their Lord, high above them, and they all do that they are commanded.
— THE QUR'AN, 16:49–50

Animals act similarly to angels. The Holy Spirit works through them, and they follow their internal guidance. Humans have free will and are arrogant. Rabbi Roller agrees that angels and animals have much in common: "Angels are often compared to animals because the character of angels is instinctive much like an animal's instinct. Angels cannot help but be what God intended for them to be."[39]

In the Hebrew written and oral traditions, angels are "spirits." The prophet Ezekiel saw a vision of angelic beings with faces of animals (Ezekiel 1:10): "Each had the face of a man in front; the four had the face of a lion on the right side, the four had the face of an ox on the left side, and the four had the face of an eagle at the back." Ezekiel 1:21 also refers to these creatures as having spirits. This vision, and that of the holy chariot and wheels, is the basis for Jewish *merkava* (chariot) mysticism. The four faces are the four directions so common in tribal and pagan religions. Rabbi Gershon Winkler teaches that "the animal image associated with each direction is more than some arbitrary symbol reflecting the attribute of that direction, but is an actual teacher/guide animal."[40] The secret of the mystics is that we are nothing but vehicles for the divine will.

The animals, like angels, willingly work as God's messengers. They show us our reflections. Rabbi Cooper points out that when one has messianic consciousness one knows that everything is interconnected. The Buddha said this, too. We forget this because long ago we ingested something — perhaps lust, desire, greed, self-centeredness, or the fruit of duality — that keeps us from seeing the true reality. Our ego and sense of separation blind us to the interconnectedness of all things.[41] If you watch the animals, you will see; ask them and they will tell you. They are also working as God's vehicles.

Faith Eliminates Prairie Dogs

> But ask the beasts and they will teach you; the birds of the air,
> they will tell you; or the plants of the earth, and they will teach you;
> and the fish of the sea will declare to you. Who among all these does not know
> that the hand of the Lord has done this?
>
> — JOB, 12:7–9

Ever since her husband, Jeff, died in a motorcycle accident, Faith prayed more often. She prayed for strength — strength to go on without the man she loved. Now she had to raise Tom, her teenage son, alone and return to work full-time at the casino dealing black jack.

The casino job was a sort of hell for Faith, not because of the men; Faith could handle men with a few choice words. It was the tobacco smoke that tormented her. Each night at the casino she inhaled enough second-hand smoke to equal ten packs of cigarettes. On the upside, the job paid well, and it kept her from being lonely. Even so, she felt antisocial, which is why she sold her ranch and moved to an isolated rural location.

This was her fresh start...a raw piece of land with a strong breeze off a large lake nearby. Like the green felt table where she dealt a winning hand, this property would unfold as a worthwhile gamble, conquering grief and assuring prosperity.

First, she fenced and cross-fenced the pastures for her three horses. Next, she put up a manufactured home. In the spring, when the ditches ran, she worked early each day digging trenches that carried water over the fields to make the grass grow. Shortly after this, her son left in search of his future. Faith felt alone. She spent her nights dealing in the casino and the days toiling at the ranch. Then the prairie dogs moved in. They reproduced exponentially — becoming too numerous to count, devouring the grass. Their burrows undermined her fence posts, and they dug craters the size of toilet bowls in the middle of the horse pastures. All her hard work fencing and irrigating flushed right down their dens.

Determined to fight for her land, she selected a firearm from her husband's gun collection. Even with the .270 caliber rifle, her aim was insufficient to eliminate the rodents. For every one she shot, she missed five, and not because she was a wimpy woman with a gun too large to handle. Prairie dogs are difficult to shoot. At the first sign of a human, they chirp warnings to each other and hide in their holes. It takes long-range weaponry for an ambush, and then one has to avoid aiming toward roadways and neighbors, making it a challenge to get a good sight on these twelve-inch moving targets. Poisons are illegal because they could kill a black-footed ferret (a protected species) by mistake. Hence, prairie dogs remain prolific in La Plata County, where ranchers have tried every technique available to eradicate them.

Faith could handle a gun; she was as tough as a tundra flower. Underneath her delicate appearance flourished a resilient, natural woman who endured the harshest weather life blew her way. She was an effeminate tomboy. Pretty without makeup, she wore jeans and a T-shirt but walked in a feminine manner. She carried her hands high at the shoulders with limp wrists, and her legs rubbed together while her hips wiggled, not in wide excursions like Marilyn Monroe's, but more of a shimmy. Her voice was feminine, too — soft and high. Whenever I vetted her horses, she talked nonstop, telling stories about her life. I examined the horses' teeth, listened to their hearts and lungs, took their temperatures, and administered dewormers and vaccines as Faith told me what happened.

Prairie dogs continued to breed like bunnies, running all over her fields. The situation seemed unmanageable. To make matters worse, Faith fell off her horse and broke her hand. Her horse Magnus was trotting along the lake when he hit soft ground and stumbled, and Faith flew off to the right, cracking a metacarpal bone. With the right hand in a cast, just the fingers poking out, life was painfully difficult. Faith needed help. One night, she finally broke down and prayed to God.

"God, I can't take it anymore. First, I lose Jeff, and then I have to suffocate every night in cigar smoke. I feel ill, like I have a cold all the time. Tom is off trying to find himself, and I have to do chores with one hand. I'm hurting, Lord! And there are all these prairie dog holes. I'm worried that the horses might get hurt falling in them. Please, God, help me. Either make my aim truer, or do something about these prairie dogs."

The next morning, Faith walked outside to have a cup of tea on the back porch and found that just twenty feet from the house was another, extra-large burrow. Faith looked up, "Thanks, God. Just when I think I can't take another thing, you throw one more obstacle at me just to show me that I can." She prepared for war.

With a .357 Magnum in her left hand and a bottle of water in the fingers of the right, she set out to wait for the beast, even if it took all day. It wasn't long before he stirred, poking his nose out. The head alone was so large that Faith screamed and dropped the water bottle, which rolled right into the hole. More determined than ever, with both hands on the revolver, she aimed down the den.

Pop. The plastic bottle flew into the air, and right behind it came a long, striped face, eyes staring into Faith's as she peered down the barrel of the gun. The monster hissed at her. She told me, "Karlene, something stayed my hand." Faith spoke to it: "You are not a prairie dog. You are...a badger." Then, wise woman that she was, she thought, "What do badgers eat? Prairie dogs! You eat prairie dogs! And when the prairie dogs are all gone, you move on." She lowered the pistol to her shoulder and backed away, thanking God. Faith was so pleased with the answer to her prayer that she took two grilled spareribs out to the badger burrow after dinner as an offering. However, by morning, she found them at the back door, uneaten.

Badger eliminated the prairie dogs, but there were still plenty of holes to fill in and other chores to do. Faith discovered that she could not repeat this story at the casino, since the men she told wanted to shoot the badger for its pelt. But she appreciated the Almighty, who did understand. It comforted her to know, as she shimmied out into the pasture each morning carrying her shovel, that someone listens; someone cares; and *that* someone is God, who works in mysterious ways.

Animals in the New Testament

Three animal symbols from the New Testament characterize the Christian God — the dove, the fish, and the lamb. Like the Holy Trinity, these animals depict three aspects of Christianity. The dove represents the Holy Spirit of the God trinity. The fish symbolizes a Christian who feeds and cares for others, like the Father. The lamb symbolizes the Savior — the sacrificial lamb offering, which God made by sending his son to die for the sins of humankind. In this manner, God eliminated the need for animal sacrifice. Jesus repeatedly said, "I desire mercy, and not sacrifice" (Matthew 9:13). According to Mark 12:28–34, Jesus wanted people to love God with all one's heart, soul, mind, and strength, and to also love one's neighbor as oneself. These two were much more important than all the burnt offerings and sacrifices.

The meaning of the dove, the fish, and the lamb symbols play out through the life of Jesus. The angel Gabriel appeared to a young unmarried woman named Mary, telling her that she would give birth to the son of God. The conception was not a sexual act and is depicted as a dove, the Holy Spirit, descending on Mary. An angel also visited the man Mary intended to marry, Joseph, and the two set off together on a journey to Bethlehem. There, Mary birthed the boy, Jesus, in a barn because there was no room in the inn. He was laid in a manger, a bunk where farm animals ate. How interesting that God let his infant son lie down with cattle, donkeys, and sheep in a barn. Indeed, this may have been the safest place, considering how animals obey God, while humans often do not. This nativity scene, common at Christmas, was first popularized using real animals by St. Francis of Assisi. He brought the ox, the ass, and the lamb to church so that people could experience what the birth of Christ was like.

The angel Gabriel also foretold the birth of John the Baptist. John grew up strong and went into the wilderness, where he ate locusts and wild honey. Then he came to Israel, where he baptized many, including Jesus. He dunked the adult Jesus into the River Jordan, and when the Nazarene came up from the water, the heavens opened and the Holy Spirit descended in the likeness of a dove and landed on his head. A voice from heaven said, "Thou art my beloved son; with thee I am well pleased" (Mark 1:11).

After Jesus was baptized, the Spirit immediately drove him out into the wilderness. He remained in the wilderness forty days, where he was tempted by Satan. He was also with the wild beasts, and the angels ministered to him (Mark 1:12–13).

When Jesus returned from the wilderness, he began his time of teaching by first gathering disciples around him. He called to men who were fishing, telling them to come with him, and he would make them "fishers of men" (Mark 1:17). He took five loaves of bread and two fish and somehow multiplied them to feed five thousand people. Perhaps these stories led to the use of the fish as a symbol of Christ. Or it may have derived from the acrostic for fish (*ichthus*) from the first letters in Greek of "Jesus Christ, God and Savior," possibly a Gnostic invention.[42] Thus the fish came to symbolize a Christian who brings others to Christ.

Written forty years after the death of Jesus, the New Testament books Matthew, Mark, Luke, and John explain that the death of Christ provides salvation through the divine spirit entering fully into the human condition and participating in suffering.[43] God became a man to be our scapegoat, to suffer so that humankind would not have to pay for its sins. Jesus was sacrificed like a lamb so that no further sacrifice would be needed. According to Christian thinking, people sinned and needed redemption. Animals did not sin and do not need a redeemer in the same way.[44]

How Do Wounds Heal?

By his wounds, you have been healed.

— 1 Peter 2:24

Shortly after speaking with the Benedictine wrangler and the padre at the Monastery of Christ in the Desert, I entered the front door of their chapel to sit and meditate. The monks were just filing out the back entrance in their hooded, black robes, each one bowing to a shrine painted with images of the last supper, angels, and saints. The top, center panel depicted a white dove, wings outstretched, surrounded in golden light. Looking up from inside the tall, octagonal tower, I could see through windows in all directions. Above and before me stood the cliffs on the edge of the Colorado Plateau rising about seven hundred to a thousand feet and exposing 200 million years of geologic history in the red, yellow, and gray-striped sedimentary layers. On the top of the tallest peak stood three crosses. The shrine sat directly below this view. On the wall to the right hung a life-size, hand-carved, wooden likeness of Jesus nailed to a cross. I was alone with these images. I began meditation with a prayer and sat in silence,

occasionally looking up at the crosses, the dove, and the crucifix, until meditation brought me into silence and a feeling of bliss.

How do wounds heal?

The unexpected question came loudly. I looked at the crucifix.

How do wounds heal, Karlene?

It seemed like a Zen koan. My mind was blank. I looked at the wounds on
Christ's body and remembered Therese Neumann, the Catholic stigmatic who
bled as Christ did from her hands, head, chest, and feet every Friday for many
years. Finally, my scientific mind kicked in with an explanation of wound healing. I thought, *Bleeding is followed by clotting, and inflammation stimulates the
production of collagen. Then cells reproduce to fill in the gap in the skin.*

The voice asked, *How do cells reproduce?*

Mitosis occurs through the replication of DNA.

How does DNA replicate?

Wow, this contemplation asked difficult questions. I thought, *Clearly, wound
healing is a miracle.* I reflected on the amazing processes in biology, such as embryology: A sperm and an egg join and develop into a human, or an elephant, or
a toad. This has always seemed miraculous to me. The process can be explained
in scientific terminology, but it does not explain how these processes know what
to do and continue to do it correctly. We take these things for granted. DNA
gets all the credit.

How does DNA replicate?

I thought about the Peruvian shamans, and the book the *Cosmic Serpent*,
which suggests a connection between consciousness and DNA. I could not remember all the steps in the process of the replication of DNA, and I realized there
would be no end to the questions of "how." So, I offered, *Christ consciousness.*

How do wounds heal?

The answer had to be the same. Christ was a healer. He touched people or
they touched him and they healed. He accepted everyone, lepers and whores.
He forgave everyone, even those who spat on him and the soldiers who sliced
his side and nailed his hands to the cross, saying, "Father, forgive them, for they
know not what they do." My heart ached as I looked at the blood dripping from
the wounds on the crucifix. With watery eyes, I submitted, *We all have so many
wounds to heal.*

How do wounds heal, Karlene?

I thought about Are Thoresen, who taught me to treat animals with only one acupoint; he calls it the "Christ Point." He received information from another stigmatized woman, Judith von Halle, who said true healing occurs only through Christ consciousness. I remembered Paramahansa Yogananda's teachings that the second coming of Christ is the resurrection of Christ consciousness in the hearts of humankind. We have access to this spiritual likeness because Christ gave us the power to become the sons of God by having faith in him (John 1:12). Christ consciousness would be a state of unconditional love, nonjudgmental forgiveness, and faith without fear, knowing the Holy Spirit is doing the healing — thy will be done.

I finally answered, *Wounds heal by Christ consciousness; our wounds heal through forgiveness, acceptance, and faith.* The peace and bliss of silence returned. *Thank you,* I added.

In hopes of becoming a better healer, I focused on Christ consciousness and noticed how difficult it is to give up judgment, to forgive, and to fearlessly trust. However, each attempt was rewarded with the feeling of bliss. I saw a lame horse that had been limping for a week and found a large screw in her hoof. Years ago, I would have judged the people as stupid for not checking the foot. In this case, I saw how kind and caring the people were. They wanted to do the right thing. Even when they ignored my advice and bought several medicines at the farm supply store instead of using what I sold them, I did not get upset. In the past, my ego would have grumbled about how irritating it is when people believe the clerk in a store or the internet over me with all my education and experience. Since my intention was to help the people and the horse, I simply showed them again how to use the hoof poultice and bandages I left them. They did not pay me at either visit. Still, I trusted, knowing compensation comes in many ways. The horse healed quickly, and I felt at peace, and payment came eventually.

Here is the process I now use for healing. Start with the intention to help. Work not for my own edification or ego, but because I care. Forgive the ignorance that created the problem. We all make mistakes and learn from them. I understand the importance of good business practice. I also have enough respect for myself to not be abused. However, I find that working to help and trust brings forth abundance. I also see that compensation comes in more ways than money. In this manner, healing happens for all.

CHAPTER 8

Science: Seeking Evidence

All too often science is presented as some collection of absolute truths,
facts set in stone.... Well that is not the science I know. The process of scientific
discovery is dynamic, frustrating, exhilarating, creative. Every day we go into the lab
and have no idea what we're going to find out. We publish our results, promote our
points of view, debate, argue, form alliances; sometimes we even create enemies.
— PROFESSOR ROBERT M. HAZEN[1]

I highlight the dynamic and argumentative nature of science to start this chapter
in order to emphasize three concepts that shape this discussion: First, because
science is ever changing, any scientific information I write about may soon become outdated. Second, because science changes so fast, science has difficulty
providing the "Truth," with a capital T, about animals or anything else. And
third, scientists study both the physical world and nonphysical energy, and this
relates to what I call the "spiritual" aspect of existence. However, drawing this
connection remains controversial, as is research into the nonphysical inner lives
of other animals — such as whether they possess emotions and consciousness.

Webster's New Universal Unabridged Dictionary (1996) defines science as
a "systematic knowledge of the physical and material world gained through
observation and experimentation." However, at this time, scientists describe
the universe as mostly empty space full of particles that form and disappear.[2]
Scientists analyze sticks and stones but also the nonphysical, immaterial realm

described by quantum physics. Quantum theory is based on concepts about the subdivision of and the transformation of energy. And yet, as we've seen, the transformation of energy is a recurring theme among world religions. In short, using different terms, science and religion might be converging in certain ways when it comes to the study of nonphysical energy.

Science is clearest and most certain when it comes to physical world history, in which Truth is often literally set in stone, such as with sedimentary deposits and fossilized remains. However, the human interpretation of these stones changes with time, and scientists find new fossils that create new interpretations of world history. The current geologic theory describes how, hundreds of millions of years ago, most of the earth's solid crust formed a single continent. Then, tectonic plate movement changed the face of the planet. In 1915, German physicist Alfred Wegener proposed the theory of continental drift, but at the time, most American geologists strongly opposed the idea. Disgraced, Wegener was laughed out of scientific circles.[3] It took fifty-five years for the theory of plate tectonics to gain orthodox acceptance. Today, geologists believe that as continents shifted, giant landmasses rose up. The Colorado Plateau, which Durango sits upon, is one such huge elevated highland that covers the Four Corners area of Arizona, Utah, Colorado, and New Mexico.

At the southeast edge of this stack of sedimentary rocks sits the Ghost Ranch. From there, one can view thousands of feet of once-buried soil from as long ago as 200 million years. Earth's story comes into view in the colorful, striped cliffs that tower a thousand feet above the canyon floor. The multi-colored stratum of yellow, red, blue-gray, pink, and green resembles a slice of layer cake raised up to the level of the frosting. The top shows the present ecosystem, and the bottom exposes 200-million-year-old soil containing dinosaur fossils. Herein lies evidence of mass extinctions and prehistoric animals, causing me to reflect on ancient mythology. There, I once observed the fossil *Vancleavea* and thought it looked like a dragon pollywog. The Ghost Ranch paleontologist agreed, describing it as a "small, aquatic dragon."

Standing before those walls, I feel diminished to a tiny speck in geologic time. Paleontologists place the earliest humans as evolving only 2.3 million years ago; animals ruled the globe for two hundred millennia before hominids arrived. At a TED talk, I once heard anthropologist Louise Leakey offer an image to describe this timeline: on a roll of toilet paper, with four hundred sheets, human beings would only occupy the last two millimeters.

The Relativity of Science

I have studied science my entire life; my father taught high school chemistry when I was in preschool. However, I only began to understand science by studying its history. I remember the moment I realized the truth about science. Stunned, I rose from the pile of embryology texts on my office floor and drove to Fort Lewis College to talk to my former undergraduate adviser and embryology professor.

"Dr. Smith," I said, rushing into his office. "I just realized that there is no such thing as a 'scientific fact.'"

He smiled, and in a soft, slow, southern accent, he said, "That's right. There are only the current theories, based on the prevailing evidence, which is constantly changing."

Do not get me wrong. As a doctor and a scientist, I am eternally grateful for all that science provides, from plumbing and penicillin to MRI technology and the World Wide Web. Even so, I am not blinded by the marvels of modern science into believing that any scientific theory is the absolute Truth. History shows us that much of what we once believed was certain is no longer true. The theory of gravity is a prime example.

"Gravity" is not a fact. Rather, the term *gravitation* is the name Isaac Newton gave to a theory that described his mathematical equations. However, Newton never claimed to know exactly what gravity is, describing it as a "force" and referring to it as "the finger of God," rather than a scientific truth.[4] Then, around 1907, Albert Einstein declared that Newton's idea of gravity was not real at all, and Einstein redefined the *theory of gravitation* as the curvature of "spacetime."[5] In other words, the effects of gravitation are ascribed to a space-time curvature instead of a force. The laws of gravity describe how things behave, and the theory explains why, but nobody knows for sure what exactly "gravity" is other than something that makes us fall down rather than up.

Yes, science provides tools that work. Thanks to Einstein's theory of general relativity, we have global positioning satellites (GPS). Of course, scientific theories work, and yet, as Professor Steven Goldman explains, "Over and over again, we discover that theories we thought were true, because they were predictably successful and gave us control and new technologies, were wrong.... So, predictive success and control are not guaranteed indices of correspondence to reality — of 'Truth' with a capital *T*."[6]

Newton came up with a theory that superseded Galilean astronomy;

Einstein's theory supplanted Newton's; and these theories will also be updated in time. Goldman writes, "All scientific theories are in a state of ceaseless revision, which raises the question of what reality 'really' is."[7] What science calls reality today is totally different from what it was in 1960. "From the history of scientific knowledge, we learn that no theory can be a fact."[8]

Furthermore, all theories incorporate a substantial number of faith statements just to operate — such as assuming that our senses are giving us reliable information, or that the telescope is accurate. All theories contain assumptions that are not themselves deducible from empirical evidence.[9]

Einstein said that science is a human construct consisting of three parts: objective reality, human imagination, and technology, where "imagination is more important than knowledge."[10] Imagination is how Einstein came up with the special theory of relativity, which states that the laws of physics are true for all observers in uniform motion, leading to the interesting conclusion that time and space are not absolute. Rather, space and time are relative to those observers.[11] This means that if your twin sister went on a spaceship into outer space, then turned around to come home, she would arrive here younger than you. Theories evolve in thoughts — imagination — first. (We make things up with our minds!) Then they become mathematical formulas, then a theory.

Veterinary school education changes constantly as well. One day, I heard from a professor in a lecture, "We used to use corticosteroids to treat laminitis, but now we know that corticosteroids can cause laminitis. So do not use corticosteroids in laminitis cases." That same day, I was on an equine ambulatory medicine rotation with three other female students, when our teacher, Dr. Schroeder, told us to go to his ranch and give his mare an injection of corticosteroids to prevent her from getting laminitis. Dr. Schroeder was nearing retirement, wore hearing aids, and was not listening to us young women as he walked away. We decided to do as he told us to; later, I sneaked into his office to leave a copy of my lecture notes on his desk. The standard of care in his time is now considered malpractice in mine. The same is true in human medicine.

In veterinary practice, we are expected to use evidence-based medicine. I find this notion frustrating because in my thirty-plus years of practice, the evidence has changed so often (and often, the evidence seems presented to promote a product) that I am not sure what to believe. Finances and politics influence science just like they influence everything else. This is one reason why I love traditional Chinese veterinary medicine; the acupuncture techniques and the herbal formulas have been effective for hundreds and even thousands of years. TCVM is ancient and existed before modern science. As scientists discover how acupuncture works, thousands of scientific studies now back up the historical results. Yet, at this time, a significant number of veterinary and medical scientists are skeptical about acupuncture.

When I began researching these topics in the mid-1990s, the ideas I had about animals having a spiritual nature were completely taboo in the scientific community, and many still are. This is true partly because science focuses on measurable data, but mainly because mainstream science has not been interested in studying the thoughts, feelings, and energetic existence of animals beyond medical care. During my research, I read about scientists who could not get funding to study animal minds because the topic was off-limits. The scientific consensus was that people were superior mentally, emotionally, and spiritually — period. Animals did not possess consciousness; they did not feel complex emotions that require self-reflection, such as love, shame, guilt, embarrassment, or pride; they did not have morals; and they did not have language. End of discussion. Anyone who dared "anthropomorphize" was shunned from the scientific community. Brave scientists such as Irene Pepperberg (*Alex and Me*) and Bernd Heinrich (*Mind of the Raven*) performed animal research against the grain.[12] No one would fund their research or publish their work. Fortunately, much has changed in the last twenty years, and research has increased regarding animal language and cognitive function. Still, only a minority of scientists admit to believing that animals are conscious or feel similar emotions to humans. Even those who believe animals are conscious rarely state it in a scientific sense because they can only make statements backed by evidence, and without funding there is little data. Furthermore, it is difficult to prove statements about nonphysical characteristics, and the researcher risks ridicule. In short, spiritual study is usually considered outside the scope of science. Or is it?

Historically, acceptance of new ideas takes time, as it did with the theory of plate tectonics. In *Kindred Spirits*, Allen Schoen describes his early veterinary years struggling against the then-accepted idea that animals feel no pain. Today, entire careers revolve around pain management for animals, and we definitely consider their emotional suffering as well. Today I read books like *Comparative Cognition: Experimental Explorations of Animal Intelligence*; *Are We Smart Enough to Know How Smart Animals Are?*; and *Beyond Words: What Animals Think and Feel*, which study animal reasoning and wonder about consciousness.[13]

Still, scientists are cautious about their conclusions. I agree with Carl Safina, who wrote *Beyond Words*, when he says that the ban on anything considered "anthropomorphic" is "bad science,...because human sensations *are* animal sensations."[14] Humans are animals.

Regarding spirituality and science, science is what made me believe in spirituality. From chemistry, I learned the first law of thermodynamics: Energy cannot be created nor destroyed in an isolated system. This led me to ask, "What happens to the energy of an animal at sudden death?" This law explains that energy only transforms into different states.

I am not alone in seeing spirituality in science; in fact, many of the early scientists considered science a spiritual pursuit. Newton, for one, believed that "God's intervention was necessary to guarantee the system's stability."[15] Geologist and paleontologist Sir John William Dawson said, "The science of the earth...invites us to be present at the origin of things, and to enter into the very worship of the Creator."[16] From the beginning, science was performed by theologians seeking God. Mathematical genius Ramanujan said, "An equation is uninteresting to me unless it expresses a thought of God."[17] God aside, astrophysicist Carl Sagan said, "Science is not only compatible with spirituality; it is a profound source of spirituality."[18] As early as the eighteenth century, when scientists were called natural philosophers, David Hume argued that matter was a metaphysical concept rather than anything real one can see.[19] If one considers the spiritual nature as the nonphysical part of being, then we must look to science, which currently suggests that only 5 percent of the universe is ordinary matter.[20] The rest of the universe is empty space occupied by what is now called "dark energy." The more I study science, the more science and spirituality blur together.

Science is fascinating and offers us a chance to explore a wonderful adventure. Let us begin with the origin of life.

The Beginning of Time and the Origin of Life

Science's creation story is the big bang theory, which is not the only model consistent with the evidence.[21] According to the big bang theory, the universe sprang into existence as a "singularity" about 13.7 billion years ago. No one knows what a singularity is or from whence it came. Upon the appearance of the singularity, it began to inflate — the big bang — from an infinitesimally small, infinitely hot, dense state, expanding and cooling. Here science leaves the conversation; what existed before the big bang and what caused the big bang is a topic for theologians and philosophers. The primary cause is not the concern of science.

Not only do we not know how the "big bang" began, the beginning of life is a similar mystery. The International Society for the Study of the Origin of Life (ISSOL) has about five hundred members, including Professor Robert M. Hazen. In his 2005 lecture series "Origins of Life," Hazen stated that scientists "do not know how life began."[22] Not only do we not know how life began, scientists disagree about how to define "life."

The scientific community has passionate and diverging opinions on the definition of life. Most researchers agree that life is a chemical system with the ability to grow, reproduce, and evolve. However, that could implicate clay as life. Clay is a crystalline substance that grows, and it flakes off pieces while passing on information from one generation to another, leading some scientists to hypothesize that clay is alive.[23] In 2007, research indicated that deep-space interplanetary dust had lifelike properties.[24] In short, no clear division exists between what is life and what is not.

Three theories indicate that life started either in a primordial ocean soup; in the deep-ocean thermal vents; or as organic molecules that landed here by way of interstellar debris — dust in the wind — and proceeded to evolve. After hundreds of millions of dollars in research, there is no evidence that a soup of molecules plus lightning or another form of energy could ever form DNA. On the other hand, the idea of spontaneous generation — life forming from nonlife — has been disproved over and over again by scientists such as Louis Pasteur. Interestingly, we have only known of DNA since 1953. Darwin proposed the theory of evolution by natural selection while having no idea what a gene was.

Today, evolution is considered an observable fact because of fossil evidence. The details of how life evolved change constantly, however. For instance, an Associated Press report on January 29, 2010, described the discovery of a fossilized footprint from a four-footed animal with a spine 12 million years before they were known to exist.[25] The theory is set in stone. Paleontologists just keep finding new stones that change the timeline.

Imagine your own beginning when a sperm and an egg unite. This form of fertilization is similar to any other one-celled creature, such as a protozoan. In humans, the cell multiplies, forming an embryo that looks like a fish. In the fish-like phase, human embryos have gill slits, a tubular heart, a notochord, and a primitive kidney. All of these persist in fish, while higher life-forms undergo further changes. Until the eighth week of development, the human embryo has a very distinct tail, and some children with "birth defects" are born with tails. The human embryonic eyes are laterally arranged, as in many other mammals, until the tenth week of life.

The brain develops similarly. To simplify the process, the human embryo forms a brain stem (found in reptiles), followed by the limbic system (seen in the first mammals), and finally a neocortex (found in primates); one part is built on top of the others like an addition to a house, such that the basement part of the human brain is an animal brain.[26] In accordance with these types of observations and distinctions, zoologists make classifications of animals.

The Kingdom Animalia

Animal taxonomy describes animals as mostly multicellular and motile, or capable of movement. So fish, frogs, insects, leeches, jellyfish, hookworms, snails, snakes, spiders, sponges, tapeworms, whales, prairie dogs, and humans are all animals. One-celled beings may be considered animals, although argument about the protozoa continues. All animals start as a single cell, so they might need to be called animals from the start; however, scientists argue about it.

Categories such as "animal" or "plant" are artificial boundaries constructed by humans for convenience in conversation and are not some label of truth. Classifications change whenever new evidence arises. Nature has some good tricks that fascinate and confuse us. For example, take the Cordyceps mushroom, called "the caterpillar fungus," which appears to transform from a worm into a plant and back to an animal again. Another interesting beast is the Portuguese man-of-war. It looks like one creature but is actually a whole group of individuals connected together working as a colony. Just as insects form colonies with divisions of labor, so do these siphonophores. Some parts swim, some gather food, and others perform reproduction.

Since all animals have life stages that are one-celled, and since some one-celled beings may be animals, and since all animals are made of cells, a bit of information on how cells live is pertinent.

Cells have organelles of respiration; they ingest oxygen and other nutrients and excrete waste. Cells reproduce and pass on genetic information via DNA.

In a larger body, they make things such as antibodies and interferon to kill bacteria and viruses that attack them. They tear down dead cells and carry them away; they build new structures such as bone. They transport things; red blood cells carry oxygen to all the other cells of a body. And cells exhibit "irritability," which is a response to stimuli used to enhance survival.

Tissues are clusters of cells, such as muscle or skin. Within the tissue, individual cells respond to stimuli and communicate among themselves by using chemical transmitters and electrical impulses in order to coordinate a response, as is done when a muscle moves. These amazing little beings manage the entire animal body, performing all the tasks of metabolism, reproduction, respiration, circulation, immunity, sensory perception — everything in the body is managed by cells, and about half of those cells are bacteria, another one-celled organism.

Humans are made of cells, and humans are considered to be conscious, but the mainstream scientific community does not accept that anything other than humans are conscious. So, my question is where does consciousness come from? If logic is followed, then it seems to me that either cells are conscious or consciousness arises somehow into a body full of unconscious parts. Or perhaps consciousness has nothing to do with the physical components we are made of but is a nonphysical, spiritual aspect of a being. In *How to Know God*, Deepak Chopra calls cells "conscious little beings." But animal behaviorists do not agree that animals are conscious, let alone an individual cell. But if only humans get that privilege, one wonders, when do humans become conscious? At conception, as a two-celled zygote? When the brain of the fetus is fully developed? At birth, or later — at two years of age, when a child becomes self-aware, or when a person graduates from college or reaches enlightenment? And why is that moment so unique?

What Makes an Animal Conscious?

In science, the term *consciousness* is the word that comes closest to the word *spiritual*. It names the nonphysical mind, our thoughts and feelings. However, I must emphasize, no one knows for certain what consciousness is. No physical substance or location in the central nervous system has been found to be the source of consciousness. It remains an enigma, something that humans and other animals lose during sleep. Yet it may depend on a healthy nervous system, since one is considered "unconscious" in a coma. However, scientists now communicate via MRI with coma patients, which further confuses the definition. Although consciousness has not been found in the brain, scientists continue to explore theories that they hope will close the gap between the physical brain and the nonphysical mind.[27]

The research that looks for the neural basis of consciousness fascinates me because the studies are done on mice, which presupposes that mice have consciousness, while animal behavioral scientists have presented no proof that mice are conscious. Professor of philosophy Daniel N. Robinson of Oxford University argues in his lecture "Consciousness and Its Implications" that since the brain (and how it works) does not vary tremendously from one species to the next, then if it is the brain that determines consciousness, any brain will do.[28]

In *Evidence of the Afterlife*, Jeffery Long studied 617 near-death experiences and concluded that consciousness is not part of the physical body for several reasons.[29] First, people who have cardiac arrest, with flatline EEG, should have no conscious experiences. However, patients remember explicit details of their resuscitations, such as which drawer in the emergency room their missing dentures were stored in, and they recall observations of the events from a perspective outside their bodies, such as from the ceiling or in a corner of the room. In other words, they are conscious when they are dead. Second, patients under general anesthesia, which is a drug-induced state of unconsciousness, also report near-death experiences with out-of-body experiences where they recall vivid details of medical personnel working on their bodies. And third, persons who are blind from birth report seeing during near-death experiences, a concept they have no framework to understand. People of all ages, all over the world, report similar experiences. Be aware that near-death experiences are not accepted as scientific by many mainstream scientists. Most believe the images people experience are caused by the brain. However, Dr. Eben Alexander explains in *Proof of Heaven* that he had a near-death experience when he was "brain-dead."[30] He claims to be able to convince any medical doctor that his brain was not responsible for his experience. Importantly, while he was in "heaven," he saw a dog.

Suffice it to say, consciousness is a nonphysical part of being, and it may correlate to some definition of "spiritual." So it warrants further examination.

According to Professor Christian de Quincey, four worldviews pertain to consciousness.[31] First is materialism, which proposes that *only matter or physical energy is real*. We already know that physics disagrees with this, and that mind (or consciousness) is not equal to the physical brain. Thoughts and feelings are not physical, therefore, de Quincey writes, "in a purely physical world, the appearance of mind would be a supernatural event." Hence, consciousness is not purely matter. Bernardo Kastrup has a nice time destroying the idea of materialism in his 2014 book *Why Materialism Is Baloney: How True Skeptics Know There Is No Death and Fathom Answers to Life, the Universe, and Everything.*

Second, dualism proposes that *both conscious mind and matter exist but are different substances.* In other words, matter is invaded by the ghost, or the God-in-man scenario.

Third, idealism says *only mind, consciousness, or spirit is ultimately real.* The world of matter is only an illusion or an emanation of spirit. There is no material stuff, only subjective perception.

All three of these views require a supernatural intervention. Enter science and the newer laws of physics to examine the perplexing relationship between the physical and the immaterial. Since quantum physics tells us that matter is equivalent to energy (as in the equation $E = mc^2$), De Quincey suggests another worldview.

The fourth worldview is what De Quincey calls *radical naturalism*, also known as *panpsychism*. In other words, since it is inconceivable that sentience or subjective consciousness emerged from objective matter, then consciousness must have always existed, and if matter is real, then it must have always been real, therefore matter is intrinsically conscious. "*Matter and psyche always go together* — all the way down."[32] So, dust in the wind is conscious, and so are cells, and all animals. Kastrup is more of an idealist and differs with de Quincey by arguing that this does not mean rocks, windmills, or computers are conscious. They do not have their own viewpoints or inner lives.[33] Science continues in the struggle to define consciousness.

Are Animals Self-Aware?

Animal behavioral scientist Marc Hauser avoids the concept of consciousness and argues that self-awareness separates humans and beasts. In 2001, Hauser wrote, "The great apes and human children over the age of two years have some understanding of self, but no other species do."[34] He defined self-awareness as awareness of one's own body and thoughts (including beliefs and desires) as being separate and different from others or one's peers. Therefore, according to Hauser, animals are not self-aware and do not have complex emotions such as guilt, embarrassment, and shame, nor do they have morals.

One criteria used to determine self-awareness is a mirror trial, where the animal being tested has a foreign mark placed on its head. When the animal looks in the mirror, rather than acting like their reflection is another animal, they touch the mark on themselves. In 2006, an elephant named "Happy" passed the test, using the mirror to check herself out, examining inside her mouth and ear.[35] Not all elephants do this, however. It then follows, with Hauser's theory, that Happy is self-aware and may have complex emotions, possibly even morals. As early as 2000, some bottle-nosed dolphins also passed the mirror test at the Osborn Laboratories of Marine Sciences at the New York Aquarium in Brooklyn.[36] It seems clear that some animals are more self-aware than we used to think.

Since children under the age of two do not exhibit this behavior, they are not considered self-aware. Scientist and idealist Bernardo Kastrup remembers the disconcerting moment he realized he was a separate being.[37] Furthermore, self-awareness fades with Alzheimer's disease. Hence, according to Daniel Robinson, "it is unclear that intuitive knowledge of one's self must be included...in examining and defining consciousness."[38]

Regarding the presence of complex emotions in other animals, consider the limbic system — the emotional center of the brain. The basic emotions of anger, surprise, anticipation, disgust, happiness, sadness, and fear exist undeniably in animals. These emotions are comparable to reactions. Embarrassment, shame, guilt, and pride are more complex and require self-reflection and self-evaluation. Love, hope, jealousy, and hate are cognitive-emotive complex emotions involving more than a single emotion. The limbic system is the "reptilian" part of the brain and is present in "lower" animals as well as humans. It is responsible for all complex emotions, from love to religious ecstasy.[39] It follows that animals with a limbic system have complex emotions. In *Unlocking the Animal Mind*, Franklin D. McMillan also argues that animals do experience feelings such as love and hope. However, the common belief in the scientific community is that "infants and animals do not share in complex emotions."[40]

One must always bear in mind that there is no way for one being to truly know what is going on in the mind of another. Even though we may not recognize consciousness when we see it, denying its presence seems premature, according to Professor Bernd Heinrich: "Dismissing the mind because one cannot personally prove it with precision by a simple test or device is like the denial of the role of inheritance before genes were elucidated."[41]

We learn new things every day. Observations of animals put to novel tests lend evidence that birds, for example, not only are much more intelligent than previously thought, but also may be conscious.

In *Mind of the Raven*, Bernd Heinrich tells of an experiment with ravens where he tied a two-and-a-half-foot string to a perch with meat hanging from the end. The only way for a raven to get the meat was to grab the string from the perch and hold it with one foot, while reeling up the string with the other. Heinrich guessed that it would take weeks of trial and error for a raven to solve the puzzle, but that is not what happened. One of the birds pecked at the string tied to the perch (a smart idea, too, since if the string fell, the meat was free). Most jumped around looking the problem over, and one bird examined the situation and performed the pull-up maneuver correctly the first try. The same bird repeated the behavior six times. Other ravens also completed the puzzle in as little as six minutes.

Heinrich defines consciousness as "the mental visualization of alternative choices that then guides judgment of new situations at the moment." The bird had to play the options through in its mind before acting; therefore, Heinrich concludes that this experiment demonstrates some level of consciousness. Furthermore, Heinrich states, "There is no evidence to suggest that humans have some new or different mysterious vital essence that other animals lack."[42]

Since ravens have long-term mates, Heinrich also suspects that ravens "fall in love" as humans do.[43] Until recently, he would have been drawn and quartered for such anthropomorphic remarks. Science only allows for statements of fact, and we can never know how ravens feel, so the idea is not scientific. However, more and more scientists are bravely stepping forward to point out that we have no proof to the contrary — that humans are the only ones with such special feelings.

One reason anthropomorphic language is not approved of in scientific research is because, as animal behavioral scientists Edward Wasserman and Thomas Zentall explain, "when we make assumptions about the similarity of our own mental states to those of other animals, this vision may be so distorted by the lens of mentalism that a clear view of the animal mind can never be gained."[44] In their book, *Comparative Cognition*, they point out that the great contribution of the science of behaviorism is to see how far science can go without the concepts of mind and consciousness. Problems arise, however, when researchers cannot hold back their amazement of the animal's behavior, as in the case of Alex, an African grey parrot.

African grey parrots are excellent speakers, which made Alex an interesting research animal for Irene Pepperberg. Alex could verbally identify objects (like a key and a ball) and colors (like red, blue, yellow). He spoke in full sentences and asked for what he wanted. And he even spelled out words, such as *nut*, phonetically. Alex could count and perform addition. In the conceptual knowledge of numbers, Alex could be compared to a five-year-old human.[45] All this he accomplished with a brain the size of a walnut.

According to a 2009 report in *Current Biology*, the size of a brain does not indicate intelligence.[46] Honeybees, for example, can count, categorize similar objects like dogs or human faces, understand "same" and "different," and distinguish symmetrical from asymmetrical shapes with a brain that weighs one milligram. Insects may be as intelligent as much larger animals. The size of the brain has more to do with the size of the body it needs to coordinate than the complexity of thought.

When Alex spilled coffee on Pepperberg's papers, which upset her, he said, "I'm sorry...I'm sorry." This is something Alex had heard others say previously, and then he added, "I'm really, really sorry."[47] It would be challenging to hear this and not believe the being speaking understands what he is saying and why; he would seem quite conscious.

Pepperberg also tested other African grey parrots with a test similar to Heinrich's string experiment, only she used a favorite bell on the end of the string instead of meat. One parrot looked down at the bell, then used his beak and feet to haul the bell up.[48] Before these tests, humans thought birds only did such things in cartoons and science fiction.

In 2012, I felt immense excitement when I read a small announcement in *The Noetic Post* about a group of scientists who declared that nonhuman animals are conscious.[49] This was such huge news for me that I had to watch the entire meeting online. On July 7, 2012, at the Francis Crick Memorial Conference in Cambridge, a gathering of prominent neuroscientists presented research on consciousness in human and nonhuman animals. They reached a number of important conclusions, including: 1) The part of the brain that humans have (the neocortex), which other animals do not have, is not responsible for consciousness. Consciousness is not a function of the brain's neocortex. 2) Self-awareness is a given. All animals are self-aware, or they would be bumping into each other. And 3) Alex, the African grey parrot, demonstrated "near human-like levels of consciousness." At the end of the conference, they published the "Cambridge Declaration on Consciousness," which states, "The weight of the evidence indicates that humans are not unique in possessing the neurologic substrates that generate consciousness. Nonhuman animals, including all mammals and birds, and many other creatures, including octopuses, all possess these neurologic substrates."[50] At last, a few scientists had the courage to state what seems obvious to me.

Scientists once thought that only humans used tools, until Jane Goodall discovered chimps using sticks and leaves as termite fishing poles. Then we decided that only humans *made* tools, but Goodall observed wild chimps making "fishing tools" to capture and eat termites.[51] Birds also bend wires to construct hooks for snagging food.[52]

We used to believe that only humans had language, and then Alex showed us how wrong we were. But even his elaborate vocalization is not considered language by some. For years some people argued that language makes one conscious. Yet there are deaf-mute humans who have no verbal language and no sign language either — no words in their minds.[53] Another example comes from humans with stroke damage in the auditory part of the brain. Words are no longer discernible; sounds are heard but no understanding of words occurs. If language makes you conscious, then these humans are not.

Temple Grandin, an autistic woman and animal scientist, was shocked to hear a professor tell her that animals were not conscious because they did not think in words. Grandin writes that she thinks in pictures (*Thinking in Pictures*, 1995), and she believes animals think in pictures, too. Does thinking in pictures mean Grandin is not conscious? Since she has authored three books, and designed one-third of the slaughterhouses in the country, and perhaps done more for food animal welfare than any human in my lifetime, I suggest that she is conscious, and so is Alex.

It seems obvious that animals talk to each other, and now science is starting to prove it. Professor Con Slobodchikoff, at Northern Arizona University, believes that prairie dogs have a more sophisticated language system than even monkeys and dolphins. Because prairie dogs have so many predators — hawks, eagles, coyotes, badgers, and humans — they have had to find "words" to describe them in order to survive. Barks, squeaks, whistles, and chirps vary depending on the approaching predator. In addition, "They have descriptive words for describing individual features of different predators," according to Slobodchikoff.[54] The prairie dogs listening to the calls react differently when the call warns of a coyote that sneaks up on them or a badger that digs into their burrows. These barks have meaning. Prairie dogs also combine old sounds to describe novel objects such as plywood silhouettes of a skunk, a coyote, and a black oval. Finally, young animals seem to learn the sounds rather than being born knowing them.[55]

In fact, if we once thought animals only acted on instinct, we now understand that animals learn, and also that animals teach one another. Humans are not the only animals who cooperate in social groups. Most animal lovers observe animals teaching each other all the time.

When Sport, my horse, was a yearling, Chief taught him about apples. Each day the horses searched under the apple trees for fallen apples. Sport seemed oblivious to what was going on until Chief snatched aggressively at an apple near Sport's foot. Chief pinned his ears like it was something to fight over, then he bit it in half and dropped part of it by Sport's feet, which he left alone for Sport to sniff, nuzzle, and finally taste.

May was our best pack mule. She was old and wise when young Sport had a one-leg hopple put on his front pastern for the first time at a high mountain camp. May was free because she never ran away. Although Sport picked on May regularly in the pasture at home, May chose to stand close to him all night long rather than with the other equines. It appeared to us that she was showing him how to move with the long rope attached to his foot, and what to do if it became tangled in his legs. He made it through the night without trouble, and she was next to him when we woke up in the morning.

Altruism is another practice considered uniquely human by some. However, even Charles Darwin recognized altruism in animals; it confused him because natural selection did not explain why individual animals sacrificed themselves for others. Social insects, such as bees and ants with sterile workers, exemplified the problem for him.[56] These guys sacrifice themselves for the colony.

The vampire bat provides an intriguing example of animal altruism. Vampire bats require a blood meal every sixty hours or face starvation. A bat in need begs of those with a stomach full of blood. The altruistic currency is blood, and giving is costly. These bats can, and do, regurgitate some blood to save lives. Crucially, this pattern of exchange occurs among individuals that are genetically unrelated. A selfish-gene perspective cannot explain this pattern. Individual bats more commonly regurgitate to those who have shared in the past. Furthermore, bats punish cheaters, rejecting requests for blood if those asking have failed to donate blood in the past.[57]

Prejudice against animals erodes away in spite of human vanity. Perhaps one day we can accept the rest of the animal kingdom as our brothers. Humans and the other animals are not as different as our appearances make it seem; rather, we all reflect the eternal beauty and intelligence of one Creation in our own peculiar way.

Translating Animal Behavior

Here are a few stories that indicate something might be going on inside non-human animals besides simple instincts. But wait! Webster's dictionary defines *instinct* as "a natural intuitive power," and *intuition* as "a direct perception of truth." "Direct from where?" I wonder. This seems like spiritual guidance to me. Of further interest, Webster's includes an obsolete definition, from 1530–40, meaning "animated by some inner force."

A woman came to Animas Animal Hospital with a fishing basket hanging over her shoulder. We entered the exam room.

"I've never liked cats," she said, "and I've seen a couple of wild cats living in the woods near our house." It was a warm day in early spring, and she described having the sliding glass door open to her deck. "I was on the phone looking out the window when I saw two cats coming toward my house. Before I could react, they came onto my deck and poked their heads through the door, dropping this."

She opened the basket, and there was a tiny, butterscotch-colored furry kitten. Under its long coat was mostly bone. I squirted a tiny amount of liquid dewormer into its mouth, and it chased after the syringe ravenously, so I opened a can of kitten milk replacement. I said, "It looks like you've got yourself a kitten." She agreed that she might have to keep it.

The kitten's parents obviously made an intentional move. They had to understand what they were doing, and they decided to leave the baby on the doorstep. This seems like a conscious, selfless act of love to me, but I will settle for "instinct."

Denise, Sue, and I were camping with our horses. We rode down to Jacob's Lake and spent the night. Sue's horses were tied to trees all night, unable to graze. In the morning, we wanted to let the horses eat for a while. Sue put hopples on the front legs of the two mares, and I stood next to Pecos while he foraged along the lake. They all had halters on, but we were not holding on to them. "When they start looking around more than they're eating, we can catch them up," I said.

But that was not what happened. All of a sudden, all three horses threw their heads in the air and off they ran, Pecos in the lead. I was inches from grabbing Sue's mare by the halter when she broke the chain between her hopples and got away. We were running after them in our pajamas. Then Pecos stopped; he seemed to realize that he was going the wrong way. He had missed the trail out of camp, which crossed a creek and climbed the steep slope to the truck and trailer. He turned and ran back past me toward camp, so I yelled for Denise to get the grain. Pecos took one look at Denise, turned, and this time headed for the trail out. I knew that if he left camp, we would have to hike all the way up the mountain and then lead them back down to get our gear. So I yelled, "PECOS!" in a way that meant he had better stop. He stopped, then he turned and trotted back to eat the grain. The mares followed, and we easily caught them.

In veterinary medicine, we call this "oral restraint." I never mistreated that horse, so he did not react out of fear. He knew what I wanted him to do, and he did it. Perhaps like a child, he just gave up the game when he realized it was not fun anymore. Whether Pecos understood my language or felt my emotions, somehow he understood and decided that eating grain was a better option than running away.

A college student brought a young cat to the animal hospital with an abscess. He did not have the funds to pay for the cat's care, so he told Dr. Carlton to "put it to sleep." I took the cat home and named him Pumpkin Pie because of his red-tabby coat coloring.

Pumpkin Pie was very timid and always disappeared if visitors came to the house. He also refused to be kept inside, relentlessly crying and clawing at the doors and windows. He loved being outside, and he came and went as he wanted for about eight years. This concerned me, since I knew there were plenty of predators, including coyotes, owls, bobcats, mountain lions, and domestic dogs.

One night Jean-Luc and I had dinner guests and were in the middle of eating when I heard the cat door open. In came Pie; he came over to the table and jumped onto my lap — a very unusual behavior for one who was always terrified of strange people. He had never jumped on my lap during meals. We were both shocked. "Wow, buddy, what's up?" I asked him as he snuggled his nose against mine while I petted him. Then he jumped down, rubbed up against Luc's leg as he stroked him, and went right back out the door. We were befuddled by this until we discovered later that he was gone. He never came back. I did not even cry because I felt like he had gone out of his way to say good-bye.

Whenever I caught Pecos in the field, I had to lead him across an old wooden bridge over the irrigation ditch. The bridge had holes in it, and the horses would look down through the holes, see the water running under the bridge, and run to get over it. As I led Pecos, the other horses would follow. One of them was Chief, a geriatric fellow with poor eyesight. Whenever I led Pecos over the bridge, he would stop and refuse to go further, no matter how I pulled on him. He would turn his head and watch until Chief made it over. Only then would he continue walking with me.

The herd is a family of individuals that watch out for one another. They care, they share, and they are aware of each individual's needs, which differ from their own. This is shown by how Chief and May taught Sport and by how Pecos watched out for Chief's safety by waiting for him to cross the bridge. That is how it appears to me, but of course, I cannot know for sure, even if it seems intuitively obvious. These are not scientific conclusions, merely observations that may be shaded by my human perspective. However, I agree with animal behavior scientist Marc Bekoff, who says, "While science has much to offer, science does not have a monopoly on truth."[58] Bekoff also suggests that animals behave as if they care for each other because cooperation, play, and fairness are important for the formation of social relationships. When animals are studied in their own worlds, they may be found to have their own form of genuine morality.

Another example of interesting behavior comes with rescue dogs. They love their jobs and get so caught up in their work that if they do not find something, they seem depressed. People have to set up fake finds to keep them happy.

We must admit there is more going on inside animals than we are currently capable of understanding. The inability to understand is not theirs but ours. Since science cannot tell us for certain, we must look elsewhere.

As Sathya Sai Baba says, "Science must confine its inquiry only to things belonging to the senses, while spiritualism transcends the senses. If you want to understand the nature of spiritual power, you can only do so through the path of spirituality and not science. What science has been able to unravel is merely a fraction of the cosmic phenomena."[59]

CHAPTER 9

Spirituality: Mystics, Clairvoyants, Channels, and Animal Communicators

When you are present, you can sense the spirit,
the one consciousness, in every creature and love it as yourself.
— ECKHART TOLLE[1]

The oneness of all existence represents the heart of spirituality. Everything is Spirit experiencing itself. Spirit resides within us and is also much more than us.

In the community of Durango, people involved with spirituality include an eclectic blend of all the religious, philosophical, and scientific beliefs we've discussed so far. Here, people participate in a variety of religious practices that might seem opposed to one another. They attend Native American sweat lodges, communicate with and express gratitude to Mother Earth, perhaps pursue Wicca, chant Hindu mantras, practice Buddhist meditation, and then pray to God on Sundays with the Reverend Michael Bernard Beckwith while watching a live-stream service from the Agape International Spiritual Center in Los Angeles. They attend the Unitarian Universalist Fellowship, Hebrew temple, and Christian churches, which support each individual's spiritual path. They watch movies on the paranormal from the Institute of Noetic Sciences; they consult shamans, clairvoyants, and animal communicators; and they learn about mysticism from people such as Caroline Myss and read *A Course in Miracles* and the Abraham-Hicks teachings, both of which are channeled material. These folks enjoy a wide perspective of spiritual beliefs and tie them all together nicely.

These beliefs and practices can also help us understand the spiritual nature of animals.

The Difference between Religion and Spirituality

Spirituality differs from religion in that religions tend to divide us, whereas spirituality aims to unite us. The main difference I see among religious beliefs is vocabulary. People of various cultures use terminology from their environment to describe the Ultimate Truth. The words are different, but the meanings are the same. We kill each other over semantics, when at the purest level, each religion teaches love.

Caroline Myss, a well-known "medical intuitive," often says, "Religion is a costume party." She believes that many ordinary people are now leaving established religions in preference for accessing God via mysticism. A mystic is someone who communicates directly with God or is aware of union with the divine. Those who experience special knowledge from God via mystical communication — such as with clairvoyance, telepathy, precognition, dreams, prayer, meditation, and channeling — find solutions from internal guidance rather than a religious dogma.

People with no firsthand experience of these spiritual abilities naturally doubt and fear them. Many people who have extrasensory perception rarely mention it for fear of ridicule. Clients tend to confess in my office. "I saw the Blessed Virgin as a child," said one. Another offered, "I have always seen dead people since I was a child." Both appeared to be normal housewives, and they both admitted they rarely told anyone else. Furthermore, the number of people having paranormal experiences is increasing. A general consensus among the spiritual community is that in the year 2012 a shift in consciousness of a large group of people occurred, raising human awareness of the spiritual nature of the universe to a new level.

Today, evidence supporting the existence of paranormal activities is found in hundreds of scientific experiments regarding telepathy, remote viewing, clairvoyance, precognition, psychokinesis, and psychic healing. One example comes from Dr. Larry Dossey, who used twenty-one double-blind studies to show that prayer works.[2] As usual, mainstream science rejects this evidence. However, spiritual believers are typically people who have already experienced the paranormal. They understand that life works on an energetic level not recognized by those measuring material results. They also envision a spiritual evolution in which the paranormal is destined to become the new normal.

Once a person experiences this nonphysical, spiritual mind state, the seed of understanding sprouts within, and he or she searches for answers among the many spiritual teachings. With spiritual growth, a person learns about spiritual laws. Some examples include these:

- What you do to others, you do to yourself — this is karma. The judgment you give is the judgment you get.
- Thoughts have power — thus, prayer works.
- We are responsible for everything that happens to us — we have to give up blaming others for our problems.
- We make choices and plans from a higher spiritual perspective — we choose our parents and have agreements to come to earth to help each other learn.
- We each create our own reality with our mind — the observer is the observed.
- Forgiveness cleans away karma — because judgments solidify our thoughts into our reality, forgiveness dissolves them.

When one accepts these laws, life takes on new meaning and becomes an inside job. However, most of us cannot accept responsibility for what happens to us. We chase money in search of happiness and struggle to manage our own thoughts. We hold grudges and blame the world for our problems. The spiritual path offers inner growth, inner happiness, and connection with divine guidance.

Most human struggles arise because of our creative and troublesome egos. By ego, I mean the notion that we are separate individuals who believe we are right and others are wrong. When human beings gained the knowledge of good and evil, or of duality, and the notion that we are separate individuals, we began to perceive the world in terms of fear for our survival. Fear created the problems of drought, pestilence, plague, and death. Humans had to kill to survive. As Eckhart Tolle puts it, "The ego created the illusion of a separate body and death."[3]

Nonhuman animals may not have the same perception or ego. They see spirit realm and know there is no death. The channel Abraham says, "The beasts do not see the birthing and death thing like humans do because they understand at all levels of their being that they are eternal consciousness."[4] According to Eckhart Tolle, animals appear to die because we believe in death. "Every being is a focal point of consciousness, and every such focal point creates its own world, although all those worlds are interconnected. There is a human world, an ant world, a dolphin world, and so on."[5] There are also highly conscious beings connected with Source Spirit, inhabiting what we might think of as heavenly realms, yet all these worlds are one.

The nonhuman animals are not confined by the negativity of an ego. Tolle writes, "No other life form on the planet knows negativity, only humans."[6] Negativity comes from our dual viewpoint. Some animals, who associate closely with humans, may become polluted by our neurotic insanity, while the others find harmony with nature, the Tao, paradise. Humans judge life through a jaded lens. Just as the ancient Chinese described the qi of the observer as being the cause of abnormal animal behavior — we are the creators of the problems we see in the world. Ouch!

Likewise, when the collective human consciousness transforms — when we find harmony with the Tao, become enlightened Buddhists, attain Christ consciousness — then the animal kingdom will reflect that transformation and, according to Isaiah 11:6, "The wolf shall dwell with the lamb, the leopard shall lie down with the kid...and little children will lead them."

The world is a reflection. If you hate it, it hates you. If you accept it, it accepts you, too. Just as the mare, Silver, moved on with her life after the death of her foal, Iris, animals teach us about acceptance.

Accepting Life as the Animals Do

If you wish to know the truth
Then hold no opinions for or against anything,
To set up what you like against what you dislike
Is the disease of the mind.
— THE THIRD PATRIARCH OF ZEN

The goal of spirituality is union with the Divine Spirit. This requires changing one's state of mind away from dualistic thinking. As we've seen, this is the goal of many religions and traditions. Union with Spirit is how shamans accomplish healing, and the way pagans view nature. It is the goal of Hindu and

Buddhist meditation practice and of the mystical practices of Muslims, Jews, and Christians. Union with Spirit is how nature and animals operate, which is what sage humans observe when in harmony with them. In order for the average human to understand what the animals already know, he or she must experience spiritual growth, which includes a number of sticky changes in perception, such as "acceptance." With acceptance, one no longer seeks to judge, control, change, or dominate others. The purpose of duality — good and evil, right and wrong — is to know ourselves. We humans define ourselves by experiencing what we choose not to be. Once we embrace acceptance, we start to see a nonlinear, nondual reality. Conflict fades away; we no longer blame others; and we surrender the ego's self-importance. We know that what happens is in our best interest, even if it hurts, because everything furthers spiritual growth. Acceptance is not apathy; rather than being judgmental, a person becomes more dedicated to resolving problems. One replaces negative emotionalized positions with the acceptance of the varied expressions of life. Acceptance brings peace, but it is not easy to accept tragedy and give up judgmental opinions. It takes immense compassion to accept that a calamity is for a higher good.

An example of how a tragedy can be beneficial comes from the death of Cecil the lion. Cecil, a Southwest African lion, lived in a national park in Zimbabwe. Oxford University scientists studied Cecil, so people kept track of him. In July 2015, a dentist from Minnesota hired hunting guides, who lured Cecil out of the park so he could shoot the lion for a trophy. Tragic, but look what happened next. The general public expressed outrage. The hunter became the hunted; because of death threats, he could not return to work for several months. The dentist experienced the terror of being stalked by predators, and perhaps he learned some valuable lessons. Furthermore, Cecil was thirteen years old. Lions only live for ten to fourteen years in the wild. Yet his dramatic death put a spotlight on trophy hunting. Money poured in to programs to protect lions; restrictions and regulations increased for trophy hunting. Today, France, Netherlands, and Britain no longer accept the import of lion trophies; more than forty airlines refuse to transport hunting trophies of any kind; and trophy hunting has declined worldwide. These changes are now referred to as "the Cecil effect."

From the perspective of spirituality, the entire event was a divine plan. The animals being hunted asked for help through the focus of their desires, and the universe provided an answer. If the purpose of duality is a way for us to learn what we do not want, then our united spirit decided that we do not want the unethical behavior of some trophy hunters, and we made changes to resolve the problem. Collectively, divine spirit, the dentist, and Cecil co-created an event that changed the world. From the perspective of higher spiritual purpose, many of us would offer to die as Cecil did, in a way that improves life on earth. Members of the military serve with this intention. Cecil was old; he would have soon lost his pride and died. Instead, he sacrificed himself for something important. Way to go, Cecil.

Things look much better in retrospect. The trick is to trust that things happen as they should. Trust is another sticky issue involved with spiritual growth. We must look for the silver lining and trust that a higher purpose exists and will be revealed in time. This is not easy, but the practice makes life less stressful.

Now, let's return to the belief of ancient shamans and pagans that animals offer themselves as food. First, look at nature. Fish eat fish, birds kill birds, mountain lions eat deer. Either we accept that the violence of nature is normal, or we accept that we are part of the turbulence we witness among animals. Paradise stories imply that humans are responsible — we corrupted nature when we lost harmony with it. Our disharmony forced us to kill to survive, and our current state of energy (qi) continues to cause the violence that now exists. We observe animals killing each other because they represent a reflection of the affairs in human society, as the ancient Chinese taught. In favor of the notion that animals do sometimes sacrifice themselves, as perhaps Cecil did, comes information from channels, clairvoyants, and animal communicators.

These spiritual teachers explain that all beings co-create reality such that when a mountain lion kills a deer, the deer is in on it. The lion asks for food, and the universe provides a deer that is willing to die. Modern beliefs contend that lions prevent the overpopulation of deer by taking out the easy targets, the old and weak. The scientific term for this balance of nature is "survival of the fittest." The channel Abraham explains that the primary purpose of beasts on this planet is to balance our energy. It follows that if we are out of balance, we hold some responsibility for the behavior of animals.

Nature shows such as *Planet Earth II* present the most amazing photography — of Tibetan snow leopards, penguins, killer whales, exotic birds, and so on — but much of the filming shows one animal being hunted by another. The vast majority of baby turtles hatch only to feed crabs and birds. Wildlife live on wildlife. Maybe this is how life must balance; or maybe, if humans became more harmonious, the pure forms of animal life found in paradise will return and the violence will end. One suggestion for resolution is to focus on beauty and love. The lion loves the deer it eats, and the deer loves to give to the lion, just as the Jataka tales (in chapter 6) describe the animals who gave their lives on the bodhisattva path. Our perception of nature colors our opinions of it. What if everything is okay and only our disturbed thoughts create the problems?

We live in a society that believes animal research benefits humankind, yet we wonder if it is humane. In 2015, the US National Institutes of Health put an end to invasive research using chimpanzees. Still, many animals live and die in research laboratories. I wondered what these animals thought, and I found a group of professional animal communicators who tackled the issue of vivisection.[7] The animal communicators were surprised by what they learned. A group of pigs explained that they felt they were doing a great service; macaque monkeys wanted to give back some of what they took in past lives. Out of altruistic love, they sacrificed themselves. Cows said, "It is our service to Mother Earth. We are all in service on this planet whether aware of it or not." A dog explained that it was a "necessary part of the soul's journey" and that each of us is on a spiritual path and may not understand the purpose of suffering. Although they appreciate those humans who demand humane treatment, these animals added that people do not help them by suffering on their behalf. In other words, our negative judgments only make matters worse — we get more of what we focus on and complain about. If we want to help eliminate animal testing, we must make energetic progress toward the good. Pray, make wise purchasing choices, clean up our thoughts, and search for alternative solutions to replace animal experiments.

Cowbirds are America's most common "brood parasite" because they destroy other birds' eggs and lay their own eggs in that nest so the other birds will raise their young. Cowbirds lay eggs in the nests of more than 220 species of birds, destroying the other birds' eggs in the process. I met a woman who hated cowbirds so much that she wanted to kill them because they threatened some endangered species of birds. I know of people who also hate cats that much because they kill birds. It seems to me that hate only hurts the hater more than anything else, and if we do create our own reality with our thoughts, then hating the animals who are simply being who they are only contributes to cowbirds and cats doing more of what they do to kill birds. Hate festers inside, growing uglier each day while not changing anything for the better. I have a full-time job just minding my own vibration let alone managing an entire species of cowbirds or trying to change the behavior of the 600 million small cats that live on earth. I can only make choices on an individual level, and I choose to find ways to make things better that do not include negative, harmful thoughts.

I have witnessed tremendous suffering, as all veterinarians do, and I have learned that I do not improve the world by carrying more pain. I have not mastered acceptance and am not even close to giving up emotionalized opinions. However, I am aware of the pain I feel with judgment and the bliss that rewards me with acceptance. These feelings constitute inner guidance. If it feels right, trust it.

In the words of Reinhold Niebuhr's famous prayer, "God, grant me the serenity to accept the things I cannot change, the courage to change the things I can, and the wisdom to know the difference."

The Quartz Lake Pack Trip

We get lost in doing, thinking, remembering, anticipating — lost in a maze
of complexity and a world of problems. Nature can show us the way home,
the way out of the prison of our own minds.
— ECKHART TOLLE[8]

Pack trips provide a connection with animals and nature that frees us from our minds. They also involve a lot of work and are often dangerous. Nevertheless, these concerns never stopped me and my friends from going because in nature we find union with Spirit. Thus it happened that three strong women, Jean-Luc, and myself set off on July 11, 2001, to camp at Quartz Lake, seventy miles east of Durango, at an elevation of around ten thousand feet.

We got an early start because, although July days begin clear and sunny, the July monsoon weather pattern means thunderstorms often roll through in mid to late afternoon. Lightning and hail at high elevations can be deadly, especially with horses wearing metal shoes on their feet and holding metal bits in their mouths. The equines had to carry all our supplies, and yet they were just as eager to go as we were. Belle, our sorrel mule, loaded into the trailer on her own before we had a halter on her.

Jean-Luc and I met the three women at the trailhead, and everyone worked for at least an hour to fill the paniers and balance the weights so the pack loads would travel safely without shifting to one side. I rode Sporty, and Jean-Luc rode Belle while leading Bonnie, a broad-chested tank of a red mule, who carried the heavy load in the solid paniers containing the kitchen and food. May, our white mule, was old at twenty-two years of age, but she was happy to go with the rest of the herd. We packed her light, filling the soft paniers with sleeping bags and clothing.

The other three women included another veterinarian, Jan, who rode my horse, Pecos, a strong, bay roan appaloosa, who was also my dearest animal friend. Jan's mother, Betty, rode a bay mare and packed a black mule. Betty's best friend, Dee, rode a wiry bay mule named Sasquatch.

Once loaded, we chose a harmonious order to avoid fights. Dee and Betty took the front, Jan and Pecos in the middle, followed by Jean-Luc riding Belle and leading Bonnie, and Sporty and I took the back, while May followed free behind us, moving at her own speed. We traveled along happily, enjoying the wildflowers and the incredible vistas of valleys and pointed peaks.

After six or seven miles, with only a couple miles remaining before we reached the lake, we turned a bend, rode out of the trees, and climbed above the timberline to a shale field. Shale rocks are chips off a mountaintop, and this shale field sat at the base of a tall, gray, north-facing peak. Trails across shale are usually flattened out enough to be safe for pack animals. In this case, however, there was no trail. Instead, the shale was covered by a huge field of snow about eighty yards wide and a couple hundred yards long. Beneath the snow, the trail would have crossed about a third of the way from the top. From that perspective, we could see the peaks in the distance all had long patches of snow on the north-facing sides, even in mid-July. The snow was hard enough for people to walk on without breaking the top, but there was no way a horse could cross it.

We were too far from the trucks and too close to the lake to turn back. After serious analysis, we decided to go down to the bottom of the snow field and climb back up around it on the other side. However, snow melting at the bottom might make for troublesome streams and bogs, and the slope on the other side was steep, about 45 degrees.

Off we went, down the slope of meadow grasses to find the streams and bogs passable. The climb up presented more of a challenge. Rock outcrops and slopes steeper than 45 degrees faced us. We each dismounted and led the animals. Sporty and even Bonnie, with her heavy load, had no trouble until about two-thirds of the way up. Both Jean-Luc and I found ourselves on rocky, slick ground trying to figure out which way to go. Then we saw the white, angel-like

May, who was above us on an easy path. She looked down at us as if to say, "This way."

"Follow May" were words we had said more than once on difficult terrain. That mule always seemed to know the best way to go, which is another reason she was allowed to go free. She maneuvered better than we did in the mountains. We followed her lead and reached the trail on the far side of the snow. I stood at the top on granite and black volcanic rocks holding the lead ropes of four animals while Jean-Luc went back to help the others. The sky was completely overcast with black-and-blue thunderheads; rain was falling at the lower elevations. The wind picked up. At any moment a storm would be upon us. I prayed for our safety and tried to remain calm.

Betty's pack mule had trouble carrying the load up such a steep grade, and the women unpacked it. Dee was the smallest person in the group and possibly the toughest. She carried the gear up the slope herself. With no time to waste, we left those packs at the edge of the snow and made haste along the carpet of alpine tundra, then descended through tall spruce and fir to the lake.

There we found a beautiful camp and unloaded "the kids," as we called our equine family members. May stayed clean at the ranch, but whenever we camped in the high country, she found the blackest mud she could and rolled in it. The black blended with the white of her coat to make the perfect camouflage. She looked like the bark of an aspen tree. We picketed each horse out to graze by attaching a leather strap around the ankle of one forelimb, attached to a rope either tied to a tree or staked into the ground with a spiral skewer. May was free. She never went anywhere and could take care of herself.

As evening rolled in, the storm rolled out, and I went back for the packs we had left at the snow field. I chose to ride Pecos because, even at twenty, he was still the strongest, fastest, most willing mount of the bunch. Most horses don't want to leave the herd. Even after a full day of riding, Pecos was ready to run. That horse loved to move fast. He walked fast and ran whenever he could. I let him pick the pace and the route. I surrendered my control to his superior skill. He had carried me over countless mountain trails, I trusted him completely, and I knew this trip would be pure joy.

We climbed out of the trees and hit the tundra at a lope. The sunset cast orange and fuchsia twilight on the dark, purple storm clouds, creating a rose-colored alpenglow on the patches of snow and gray peaks. Clear, calm tundra pools echoed the sky's reflection. The sound of hoofbeats and my beloved horse's breath kept rhythm with my heart. On top of the world, where heaven and earth illuminate each other, my mind was vacant. Here all the righteous dogma and intellectual opinions fall like chips of shale off a mountain. Free of the confines of clocks, phones, computers, and their associated mental flotsam, I was empty — a conduit for Spirit, at one with the nature of the beast who flew in harmony with Mother Earth. This is spirituality. Surrender, accept, trust, love, joy, ecstasy, peace, bliss. Money does not buy it; it does not come in a bottle. If I could only have one memory, it would be riding Pecos on some high mountain trail feeling this unity with Spirit.

I tied the packs behind the Australian saddle, and we ran back to camp in time to help with dinner. Jean-Luc had the tarp strung up and sleeping bags fluffed. The women told stories, and we laughed; we enjoyed the amazing array of celestial lights visible at that altitude and slept well.

We spent the days hiking, relaxing, and doing daily chores of cooking, cleaning, and caring for the equines. The mules could stay in one place for several days, but Sporty annihilated the vegetation beneath him in half a day. Usually he managed to get his picket rope wrapped around some poor bush like a spool of thread. He would be snubbed up, tight against it, looking at me, nickering, "Get over here and untangle me." I would unspool the rope, and he would move on without showing the slightest hint of appreciation. Even on day rides, when we would stop for lunch and I would tie him up short and high, that horse would tangle his neck in the lead rope. He was always busy, destructive to himself, so I had to keep one eye and ear open for him at all times.

Often, I considered how my horses reflect my self-image. Pecos was pure pleasure, but Sport showed me another side. Was I too busy, too frantic, harming myself? Did I not appreciate all the loving assistance Spirit gave me? Yes, I was just as obstinate as Sport sometimes.

One night, thinking about the ride home, I realized that it would not be safe to take pack animals down the way we had come, around the snow field. I brought it up at dinner. After some discussion, we decided to shovel a trail through the snow field before starting our trip home.

Our only tools for digging were one small shovel and three metal pie pans we used for plates. I offered to fill the feed bags for the horses with dirt to spread on the trail. That morning, we packed lunches and rode off. When we arrived, we tied the equines to some cinquefoil bushes near a tundra pond, just below the shale field, and went to work. It was a beautiful morning, and we made a decent dirt-covered trail across the entire eighty-yard snow bank. We never did find the bottom of the snow, which was deeper than we could dig with our pie pans.

When we returned to the horses for lunch, we found several were loose, including Sporty. Sport was teasing the bay mare, kissing her muzzle; everyone was having a lovely time together. We ate lunch, and Jean-Luc and I cooled off in the pond, then we went back to camp for a final evening in heaven.

The day of our return ride, I felt worried — a useless emotion that only causes more problems. I knew the people would be safe. The animals were my concern. I prayed, asking for them to all have a safe trip home.

The three women headed out before us. As Jean-Luc and I traveled the tundra trail, I could not contain my fear, knowing it would only cause trouble. My inner guidance said, "All is well," but I had difficulty believing it, so I asked for a sign. Within moments, I looked down at the trail below me and saw movement. Camouflaged ptarmigan chicks, the color of the lichen-covered rocks, scurried off the trail from beneath our mules and Sport — one, two, three, four, five, six, seven, and a mother hen, making eight tiny birds. We had eight equines, so I took it as a sign. All was well.

When Jean-Luc and I arrived at the snow, the others had already led their horses across safely. Betty hiked back to help lead ours. We sent May off first, on her own. She gently and slowly walked out to the center, then sunk up to her belly in the snow. She rolled to her right side to get out, and it looked as though she would roll off the edge. If that happened, the green poncho covering her paniers would make her slip and slide on her side all the way down the snow to the rocks below. We all gasped as Jean-Luc ran to help her, but before he arrived, she righted herself and made it to the other side.

The other two mules were anxious now, and Bonnie broke away from us and barreled through the snow toward Jean-Luc, who had to jump off the side of the trail to avoid being knocked down by her plastic paniers. She looked like a snowplow on a mountain pass, snow flying everywhere as she made it across. Belle was braying and bouncing in place. Betty let her go, and she too made the trip safely.

It was Sporty's turn, and he wanted to go. He squealed like a little girl and bucked in place. I tied the reins together, looped them around the saddle horn, and let him go. He ran out to the center of the trail, which by now was the consistency of mashed potatoes. He stopped in the mush and noticed the other horses were not straight across but farther downhill. I could see him consider the situation and decide that he did not want to follow the trail; he wanted to go down to the horses, so he leaped off the edge of the trail onto the snowfield. "Sporty!" I yelled, with mournful concern. He squatted down low on his hind limbs, spread his forelimbs out long and wide in front of him, and skied down the slope. I started laughing. Betty said, "How can you laugh!?"

"Look at him. He's having fun," I replied. His facial expression looked like that of a little child joyful at play. Sporty glissaded down the slope about forty yards to just above the elevation of the others, and then he made a lovely stem christie ski turn to the left, glided over to the grass, and trotted off to join his herd.

Sport was connected to Source, in harmony with the Tao, and guided by God. He made fun of my worry — he made sport of my concerns. Sporty taught me that day that if I can find ways to surrender, accept, trust, and focus on joy and love, all the terrors of life can transform into a playground in paradise.

Animal Communicators

Interspecies communication happens all the time with one animal understanding another. This is classically defined as the transfer of information from a sender to a receiver that affects the behavior of the receiver. A cat hisses at a dog, and the dog backs off. Humans and animals often understand each other's vocalizations and subtle postural cues. As I have mentioned, animals also sense people's energy. Another level of interspecies communication comes from people who use psychic abilities to transmit information. The psychic sends and receives information outside the normal range of the senses by way of clairsentience (feeling), clairvoyance (seeing), clairalience (sensing odors), and clairaudience (hearing). Many of the people I know who perform psychic animal communication are veterinarians. Most do not talk about it for fear of ridicule.

It seems to me that every veterinarian has advanced animal communication skills, and most people who love their pets do as well. However, maybe we give ourselves too much credit and should instead attribute the communication skill to the animals. For example, one day I walked into the back door of Animas Animal Hospital and found a yellow Labrador lying on the floor next to the door. He looked up at me and made one intense, high-pitched whine — "Ueehh."

"I think this dog needs to pee," I said, looking around the room at the technicians and doctors, who were all busy with other patients. I offered to help and learned the dog had become acutely paralyzed from the neck down when he jumped a ditch and hit the opposite bank. A technician held the dog as I palpated the bladder. It was as big as a volleyball and hard; the paralyzed bladder would not express. The client planned euthanasia for the dog, but as divine intervention would have it, I arrived just in time. Dr. Albert used a urinary catheter to relieve the bladder, and I performed electro-acupuncture, which I offered to try free of charge to see if it might help. I treated the dog, and afterward, the client called to say the dog was better and to ask me to treat the dog again. We were all amazed when, immediately after the second treatment, the dog walked on all four limbs. Hurray for acupuncture, but my point is this: One whine told me the dog had to urinate. I was not the communicator; the dog was! Our pets have figured out how to tell us things, and we understand. Veterinarians have to learn the nonverbal languages of many different animals — a look in the eyes, the flinch to a touch, a tail flick, and the set of the ears add up like words and phrases. Sometimes we understand more than just physical cues and common vocalizations.

As stories in this book illustrate, I talk to the animals as if they understand, and they sure seem to. My clairvoyant friend Dana says animals "read our pictures." When we speak, we form an image in our minds, and the animals see that and know what we mean. The animals are clairvoyant, too, and this is one way animal communication occurs.

One day I heard a dog speak to me. She was a new patient, and I knew nothing about her. The client came into the office and immediately left the dog with me while she visited the restroom. The dog sat down in front of me, wagging her tail and looking me in the eyes. I said, "Well, hello. How are you?"

The dog appeared healthy, but in my head I heard her say, *Oh, not so good.* She looked away.

"What's the matter?" I asked.

My stomach hurts and I don't feel well, she said.

"Maybe we can help you," I said.

I don't think so, she answered, *I think I'm dying.*

When the woman returned, I learned that the dog had kidney failure from eating melamine-tainted dog food. She died the next week.

When I began studying animal communication in 2011, I took an apprenticeship with Kate Solisti. I also read several books by other animal communicators, including *The Language of Miracles* by Amelia Kinkade. Following her guidelines, I made my first attempt at intentional animal communication for a client named Sally, who called to ask me why her dog, Mandy, might suddenly develop separation anxiety. Sally returned home from vacation, and the dog seemed normal, and then Sally left again to go grocery shopping. She returned an hour later to find the door frame chewed and clawed away. Ever since then, Mandy trembled and cowered in fear at the prospect of being left alone in the house. Mandy was a middle-aged, spayed female who was always calm and gentle. The behavior was certainly out of character, and Sally suspected something frightened her. While looking into the eyes of a photo of Mandy on a computer, I said a prayer, imagined a bridge of light between my heart and Mandy's, and I asked the following questions.

> KARLENE: *What is your favorite food?*
> MANDY: *Chicken, baked chicken meat; not fish.*

Years earlier, I had prescribed fish for Mandy, when she had red, hot rashes.

> K: *What is your favorite treat?*
> M: *Meat, chicken.*
> K: *What is your favorite toy?*

I saw an image of a dumbbell-shaped, red-and-white cloth toy.

> K: *Are there any other dogs in your house? Are there any cats? Do you have any friends?*
> M: *No. No. No.*
> K: *Do you want a cat?*
> M: *NO!*
> K: *Do you want a dog?*
> M: *Yes, I want a puppy, a little dog. I like little dogs.*
> K: *What makes you so upset about being in the house alone? Did something scare you?*
> M: *No, I'm bored and lonely.*
> K: *Do you get enough exercise? What do you like to do?*
> M: *No. I want to explore in the woods. We never get to do that anymore.*
> K: *Do you have a job?*
> M: *I guard the house. And I'm sick of it — I want to get out of here.*

Amelia Kinkade suggests that if the animal expresses some emotion, the communicator should attempt to feel that emotion, which I did, and it felt horrible to be so locked up that I wanted to tear the door off.

> K: *What about your guardian annoys you?*
> M: *She ignores me.*

At this point in the conversation, Sally and her husband entered the room. I ended my talk with Mandy and told them what I heard. Sally had switched the diet to chicken recently and Mandy loved it, preferring it to the white fish I had prescribed previously. The toy was a mystery. It became very clear, though, that Mandy indeed did not get enough exercise (she only received a thirty-minute leash walk once a day). She was afraid of cats, had no playmates, and loved little dogs. I learned that Mandy was left at home a lot. It fit that what scared Mandy was the idea of being all alone, locked in the house with nothing to do. Sally admitted that she punished Mandy for her bad behavior by ignoring her when she came home instead of greeting her warmly.

"We are not getting another dog!" Sally's husband said.

"Maybe she wants to go to the dog park," said Sally.

"She said she wants to explore in the woods," I repeated, as I knew the dog park had only a few scattered trees. I felt ill. It was clear that Mandy would not be taken to the woods, she would not get more exercise, and she was not getting a companion. I went to bed feeling defeated and depressed. All I could do was pray for Mandy to be relieved of suffering somehow.

A few days later Sally called. A friend had acquired a cairn terrier, and Mandy spent an entire day at her house playing with the dog. They loved each other and had such a good time that the little terrier dog came to play at Sally's the next day. The prayer was answered. Four days later, while on a walk, Mandy went into a bush and came out with a dog toy matching the description of the one I saw. It was a red-and-white rope with knots on both ends.

It seemed like I had indeed communicated. However, I did not like feeling another animal's pain. After twenty-six years of veterinary medicine, I had witnessed enough suffering. A psychologist had already told me that I had post-traumatic stress disorder from working long hours after twenty-four years of emergency work. When I discussed Mandy's case with my psychic friend

Dana, she said, "That is why I never practice clairsentience; I don't want to feel everyone else's pain."

CLAIRSENTIENCE

Clairsentience is one of several psychic abilities people use to communicate with animals. It is the empathetic ability to feel the emotions of others. This is what happened when I communicated with Mandy. It is also the ability to sense the energy surrounding something or to feel something beyond the five senses. Some people can sense heat radiating from an animal or feel how a medication affects an animal's energy. Clients come into my office with these skills on a regular basis. We often work together to determine the best treatment plan. Clairsentient abilities can also become learned skills for veterinarians and animal handlers who may not think of themselves as psychics. People with pets can feel when something is wrong. I always listen to clients who say they sense something is wrong with their pet even when there are no outward signs.

I learned to feel pulses as a part of the diagnostic process of traditional Chinese veterinary medicine. TCVM is prescience. Veterinarians long ago had to make a diagnosis without the help of X-rays, blood chemistry values, thermometers, or stethoscopes. The ancient Chinese taught three main techniques to make a diagnosis: They looked at the tongue, which is like looking at an organ because there is no skin on it, in order to get an idea about the circulation of blood and energy to the organs. Then, they felt the acupoints to sense tenderness, which indicated the areas of the body that were out of balance. And finally they felt the pulses to understand what was happening with the circulation of qi and blood.

Pulse diagnosis seemed impossible to learn at first. Sure, I could feel the pulse of the femoral artery on the hind limbs, but it was beyond my ability to differentiate qualities like weak (indicating a blood deficiency), wiry (indicating stagnation of blood), slippery (indicating accumulation of dampness), thin (indicating yin deficiency), fast (indicating heat or fever), and so on. Plus, the animals did not stand still, and their fat and fur got in the way. Then, a teacher taught me to use an energetic pulse technique. With this method, I feel my own pulses at my wrist. There are twelve positions to feel, which correspond to each of the organs of the body. As I feel my various pulses, I then touch an animal and see how my pulses change. For example, I feel my liver pulse and then touch an animal and find that my pulse gets weaker when I do. That tells me the liver energy of the animal is weak. When I first learned this, to test my accuracy, I examined each patient and made a pulse diagnosis before my teacher did. I got the same diagnosis she did in every case. From then on, I practiced both femoral and energetic pulse taking, eventually becoming confidant with both.

One day in the reception area at Animas Animal Hospital, a woman I knew was crying and holding a cat. I asked what was wrong, and she told me that she was there to have the cat put to sleep. She said she could not bear to hold her during the procedure and asked me if I would do it instead. I agreed and carried the cat into an exam room, where I met Dr. Albert. He took one look at the skinny, geriatric feline and said, "Chronic renal failure." A fair guess, since most old cats die that way.

I checked the energetic pulses and said, "The kidney pulses feel good to me."

As this was early in my relationship with Dr. Albert, he rolled his eyes and said, "Then what's wrong with her?"

"The heart pulse is nonexistent," I said.

Dr. Albert grabbed a stethoscope and looked surprised. "The heart is in atrial fibrillation." He read the record and found that, years earlier, the previous owner of the clinic had diagnosed congestive heart failure.

Suffice it to say that, after numerous cases like this, I trust the energetic pulse technique. This is an example of learned clairsentience, although it does not seem paranormal to me. I compare the sensing of energy to the way a dog can tell if they like someone or not; dogs know immediately who they like and who they do not. They may get more information from their sense of smell, but Caesar Milan, the dog whisperer, explains how dogs feel our emotions, which is why he encourages people to be calm, confident leaders. Research shows that dogs read human emotions, which I imagine seems pretty obvious to pet-loving people.[9] Research also shows that the heartbeats of dogs and their owners sync up when they reunite.[10] Animals sense energy. They see and feel things we ignore. This knowing is normal for animals. It is normal for humans as well, but human animals have gotten lazy about using extrasensory perception, such as clairsentience.

One day, I was performing moxibustion on some acupuncture needles in a horse's back. That is, I was using a burning herbal, cigar-like stick to heat the needles. Typically, when the needle gets hot, a horse will move away, twitch her skin, or swish her tail, but this horse just stood there for much longer than normal. As I talked over her back to the owner, the entire horse suddenly jumped in place. She jolted like a person who lurches awake when falling in a dream. As I lifted my left hand from her withers and my right hand with the moxa stick from her lower back, a bolt of energy surged between my hands, back and forth. I had heard people could feel energy off the needles, but I never expected to. It surprised me to feel energy.

My empathy for the dog Mandy and the sensations I feel from pulses and needles are examples of sensing beyond what may be considered the norm. Veterinarians use their sense of touch and intuition more than medical doctors, whose patients can speak. Furthermore, animal doctors do not have carte blanche to run every diagnostic test needed. We have to develop skills at guessing, for lack of a more scientific term. Intuition is the best word to describe how we first start to make a diagnosis. Then we perform the most cost-effective tests to back up our hunch. Clearly, extrasensory perception is useful, and perhaps veterinarians use it more than they realize.

CLAIRVOYANCE

Clairvoyance means seeing beyond the range of normal vision. I first learned the energetic pulse technique from a veterinarian in Florida who was a student of a Norwegian veterinarian, Are Thoresen. Are has cured animals and humans of cancer using a one-acupoint technique. I read his scientific papers and began using his diagnostic and therapeutic one-needle treatment for each of my animal cancer cases about eight years ago, and I have had very good results. I also used Chinese herbal medicines, nutritional supplements, and dietary changes. These results intrigued me, and I decided to study with Are in April 2013. I attended an International Veterinary Acupuncture Society class in Banff, Canada, along with twenty-one Canadian veterinarians. I was the only veterinarian there who had used Are's pulse diagnosis technique before.

When Are started teaching the class, he told us to look into the heart of the animals and feel the change in our own pulses. "Look into the heart? What, are you clairvoyant?" I asked. It turns out that Are is a salty, old, clairvoyant Viking who resembles Gandalf — tall with long gray hair.

"Yes, you look at the heart, then go inside and see the blood moving around and feel your pulse. That is the spiritual pulse," Are explained.

I was used to feeling the animal while checking my own pulse, and I found it difficult to sort out all the energy of the other students and the people who brought the animals. "Just block that out," said Are.

That seemed impossible, and I found this new way difficult. Then we examined some horses. One was a four-year-old, black-and-white paint mare. A little girl was riding her bareback, and the horse was very lame in the left hind limb. She hiked her left hip and dragged the left hind toe. Several veterinarians had treated the horse previously with medications, chiropractic, and acupuncture, and the horse had not improved. Are made the diagnosis of a kidney deficiency and used a therapeutic laser on the point between the left hind heel bulbs. He also used it on one point on the right side of the sacrum and pulled on her tail for a couple of minutes. Then we moved on to examine the next horse. About twenty minutes later, I turned to see the little girl trotting around on the mare. The girl was smiling and nodding her head, saying, "She's better." She was dramatically better and went sound and stayed sound, according to the veterinarian who examined her a month later.

That was enough for me. I wanted to do what Are did. In the classroom, we treated dogs. I still struggled to keep people's energies out of my focus. Are said that a golden retriever had a heart deficiency, and I could not feel it. "Try again," he said. I focused and felt again, but my heart pulse did not weaken when I looked into the dog's heart. Then the dog's woman petted the dog, and my heart pulse sunk.

"When she touches the dog, I feel it," I told Are.

Are looked at the woman in the chest, then turned to me and said, "You're right." Then he asked the woman, "What happened to you three years ago?" (Are could determine how long ago a problem began, a technique he taught us.)

She said she had broken up with one boyfriend, got another one, and started a new job as a psychologist, counseling abused children. "It's you," Are said. Are believes that dogs take on a lot of their human's energetic patterns. I have witnessed the same. However, not all problems come from the people; animals have their own issues, too, but those who live together can sense the energies of the rest, and they do affect the entire household. Plus, we have to consider that the dog's person may have been perceiving a problem in the dog that was indeed their own.

Some online dictionaries define *clairvoyance* as the supposed ability to read the future, but no reputable clairvoyant I have ever met would agree with that. All I have ever been told is that no one can predict the future for certain. One can only see what things look like now, and even if the evidence strongly suggests a negative future, the clairvoyant must find some way to be helpful. "Words are powerful — be careful what you say," explains my friend Dana. My advice for anyone looking for a psychic reader or animal communicator is to find a person who focuses on positive outcomes. Avoid people who give negative readings.

CLAIRALIENCE

Clairalience is the ability to smell things not actually present. People are said to smell cookies or baked bread when the archangel Michael is near. An episode of clairalience occurred for me during an animal communication with a dog that had chronic eosinophilic pneumonopathy, a lung disease that was causing congestion and thick nasal discharge. I received a distinctly rank odor associated with the nests of pack rats that the client had found under the bathtub and pantry in her old home. It smelled like a combination of feces, urine, and garbage. A dog's sense of smell far exceeds a human's.

One of my favorite clients, who became a dear friend, had a dog that suffered from sinus problems for years until he died. When her other dog also showed up with a thick, nasal discharge, I urged her to check her home. She did and found a dangerously high radon level. Mitigation took some time, and about a year later, my friend died from lung cancer. From this and other experiences with sinus problems in dogs, I recommend that anyone whose dog has a chronic respiratory disease have the house tested for radon, mold, and other infestations such as pack rats.

CLAIRAUDIENCE

Clairaudience is the ability to hear sounds not audible by the normal ear, receiving thoughts or messages mentally. This is what happened for me when the dog came into my office that told me she thought she was dying. It is a common technique used by animal communicators. Although many scientists deny this sort of thing happens, I know a number of well-respected veterinarians who use clairvoyance, clairaudience, clairsentience, clairalience, and animal communication. They do not talk about it or share their stories in public discourse like I am here because of the risk of ridicule from colleagues. Many veterinarians, including those I studied with in Canada, who would not consider themselves psychics, have learned to use Are's pulse diagnosis as well. Many clients hire animal communicators and feel free to share the information with me, while not speaking about it with other veterinarians whom they know would laugh at them. I am amazed at the number of clients who bring animals to me who already feel comfortable with my esoteric practices; they actually prefer them, after having personal experiences themselves with what might be considered paranormal. Often clients tell me they already believed in such things before they met me. Even old ranch folk care little what I do as long as the animal gets better. Since my practice has thrived for many years, I feel comfortable sharing. Further, I hope veterinarians become more respectful of their client's beliefs. It is good for business, and we all learn when we open our hearts.

Consciousness and Spirituality

Scientists have shown that brain activity and brain structure change with meditation.[11] The practice of spiritual endeavors changes the brain function and body physiology, and it establishes a specific area for spiritual information in the right brain prefrontal cortex and its concordant etheric (energy) brain.[12] I have focused on spiritual teachings for twenty years, and I have frequently felt bliss during meditation. Over time, I began to view the world through a more accepting lens and to understand that this is what people call "spiritual growth." However, this change has affected not just my frame of mind; my brain has actually changed.

Jill Bolte Taylor, a neuroanatomist, experienced firsthand how the brain works when she had a stroke that damaged her brain's left hemisphere. She discovered that we are able to have a mystical experience based on our brain anatomy. With her left brain temporarily turned off, she realized that everything is energy. Our sense of hearing is an example. As wavelengths of energy beat upon our eardrums, the organ of Corti part of the inner ear translates the neural code. The left hemisphere organizes the energy patterns and interprets them. Without the left brain, sounds are not separate. Jill was not deaf, but she could not distinguish words; she could see, but not differentiate colors or boundaries between objects. She wrote, "Although most of us are rarely aware of it, our sensory receptors are designed to detect information at the energy level.... You and I are literally swimming in a turbulent sea of electromagnetic fields."[13] During her stroke she experienced life without the dualistic left brain: "Everything radiates pure energy; your heart soars in peace as you swim in a sea of euphoria." Brain bliss is in your right mind, according to Jill: "The right brain is the seat of the divine mind, intuition, and higher consciousness."

Jill's experience confirms my position on consciousness in animals. She could not speak or understand words, yet remained conscious; consciousness does not depend on language. Without the dualistic left brain, she realized we are all one, including animals and plants — everything. She was also aware of the "consciousness of the cells" making up her body.[14] Cells are conscious.

In his book *Transcending the Levels of Consciousness: The Stairway to Enlightenment*, David R. Hawkins, a psychiatrist and spiritual teacher, defines consciousness as an all-pervasive universal energy field of infinite power and dimension, one that is beyond time and is compositionally nonlinear.[15] The not-manifested, nonlinear creator is capable of knowing itself solely by virtue of consciousness, which is the matrix of existence.

Hawkins measured consciousness levels using the relative strength of muscle testing. This clinical science works through the nervous system and the acupuncture energy system, and it functions similarly to Are's pulse technique. With truth, the musculature is strong against pressure because the acupuncture channels become strong, and with falsehood, muscles go weak because the acupuncture channels get weak.[16] Many of my clients have experienced health benefits from health-care practitioners who use muscle testing, also known as applied kinesiology. These people bring their animals to me because I use a similar technique, pulse diagnosis, to determine which foods and medications are beneficial and which are not. By the method of muscle testing, Dr. Hawkins charted the levels of consciousness from 1 to 1,000, with 1 being bacteria and 1,000 being the great avatars — Jesus Christ, Buddha, Zoroaster, and Krishna, and the higher levels of enlightenment. He defines the condition of enlightenment as the experience of divinity within as self or identity with God immanent.[17]

In his book, Hawkins compares the levels of human consciousness and the levels attained by animals, and he lists some nonhuman animals as possessing higher consciousness levels than some humans. This feels true to me. Rather than saying that all humans are conscious and all animals are not, we have to admit that we observe a spectrum of awareness among beings and at various times. Hawkins measured 80 percent of the world's human population below the consciousness level of 200, whereas he places the average deer at 205. Our narcissistic ego locks humans in primitive self-interest at the lower consciousness levels. As consciousness evolves over time, some animal species, as well as portions of humankind, rise to consciousness level 200. Brain physiology changes dramatically at consciousness level 200 "in both man and animal" from predatory to benign. There is also a difference between experiencing and awareness.

Below consciousness level 200, the animal or human experiences the life process, but there is no conscious awareness of existence. So, a frog or average human experiences life but is largely unaware of anything but the basic survival instincts of personal gain. The cat's purr, a dog's wagging tail, and a bird's song all measure at the level of love, or 500. Hawkins states, "Love also appears in the animal kingdom as an accompaniment to the wagging tail of the dog and the purr of the cat, and it also expresses as the maternal love in its self-sacrificial nature."[18] Hawkins placed Alex the African grey parrot at the level of reason, or 401.

It is important to understand, Hawkins notes, that different levels of consciousness do not mean that one is better than another but rather that each is different. That sort of dualistic thinking, distinguishing better and worse, is what one transcends on the way to love. Hawkins does not see the ego as the enemy, rather more like a puppy, making mistakes but forgiven for its innocence. In reality, there is no separation of this or that. Instead, one becomes unconditionally compassionate and forgiving to all life in all its expressions, choosing to see the beauty, perfection, and sacredness of all life.[19] As one becomes aware, aligning more with love, the more love one sees. Elsewhere, Hawkins writes: "The universe responds to love by revealing its presence.... Divinity shines forth from all Creation to those who can see. Nature becomes not unlike a children's cartoon where the trees smile, animals talk, and flowers move gaily."[20]

Although Hawkins was a psychiatrist by profession, his writings speak of mystical experience, and he calibrates the truth of the mystical teachings at the level of 600, that of "Peace, Bliss, and Illumination" or above.

Mystics

Mysticism is present in all major religious traditions. Mysticism is the direct intuitive experience of God, a union that is accompanied by bliss. The mystics view God as both transcendent and immanent, an interpretation known as "pantheism." The Jewish mystics are the Kabbalists, with doctrines that include the transmigration of souls and the wisdom that everything is God.[21] As Rabbi Gerson Winkler writes, "All creation is imbued with consciousness." And also: "All plants, minerals, and animals, including stars, moons, suns, and planets, are living, conscious beings replete with divine wisdom and soul."[22]

The Hindu yogi attains connection with the Atman, the higher self, breath, or vital force, which is in every life-form. The Sikhs attain relation with God via the Atman, as God abides in the Atman and the Atman abides in God.[23]

The Islamic Sufi seeks close, direct, personal experience of God in order to know absolute reality. As the Sufi poet Rumi states, "That which is false troubles the heart, but Truth brings joyous tranquility."[24] One must have absolute trust and obedience to God. The Sufi uses techniques capable of producing trance states of ecstasy, such as whirling. John Bowker writes, "There is nothing but God, at the same time, there is also a transcendent otherness of God."[25]

The shaman's ecstatic experience also denotes a realm of bliss accompanied by mystical experience, where the material world transforms for the spiritual adept.

Buddhists describe mystical states and truth, yet they strongly contest the term Atman or God, since any word used to define such an idea is a mental construct and another illusion.

Reverend Michael Bernard Beckwith is another mystic whose teachings go beyond the category of "Christian." Mystics find no boundary between anyone and God, since we are all "thought-creations" of Spirit. Beckwith writes, "Everything in the three-dimensional world is made of Spirit's thought vibrations condensed into form."[26] There is no dogma or doctrine for the mystic. Their union with the divine gives them insight and often extraordinary sight or clairvoyance.

Christian mysticism, most common among the Catholic faith, also seeks perfection in austerity and silence. As the monk Thomas Merton wrote:

> Let me seek, then, the gift of silence,
> And poverty, and solitude
> Where everything I touch is turned into prayer;
> Where the sky is my prayer, the birds are my prayer,
> The wind in the trees is my prayer,
> For God is all in all.[27]

A Happy Ending

Death is a mirror in which the entire meaning of life is reflected.
Death is the beginning of another chapter of life.
— SOGYAL RINPOCHE, *The Tibetan Book of Living and Dying*[1]

Days before his hundredth birthday, my great-grandfather went to bed one night, fell asleep, and never woke up. Most of us would define that as a good death. However, a warrior may prefer to die in battle for a righteous cause, which tells us that the definition of a good death differs for each individual. Cats like to hide and be alone. Dogs seem to prefer being held by loved ones. The animals, who have short lives, teach about impermanence. They remind us that the most important lesson death teaches is to appreciate life.

We Americans live in a society that in general approves of capital punishment — an eye-for-an-eye murder — yet also disapproves of "death with dignity," or assisted suicide for loved ones who suffer with terminal illnesses. At the same time, many view someone as cruel if he or she does not euthanize an old pet in a timely manner. These conflicting ideas about death result in confusion and fear. The number-one fear people come to my office with is the fear of doing the wrong thing at the end of a pet's life. Everyone wants a good death.

With the intention to calm fear and clear up confusion, this book's final chapter examines spiritual concerns about the moment of transition, and I offer some insight into how to improve that event for person and beast. Since we know the energetic essence of a being lives on, I refer to death using the term *transition*.

The Good Transition

What dies is the ordinary mind and its delusions.
— SOGYAL RINPOCHE, *The Tibetan Book of Living and Dying*

The inner science practiced by Tibetan Buddhists represents a comprehensive knowledge of the mind. From their view of the enlightened mind, observed over many centuries, comes an explanation of what happens at death. These teachings are compiled in *The Tibetan Book of Living and Dying* by Sogyal Rinpoche. The author offers advice to people who seek enlightenment: the escape from the cycle of rebirth into the world of suffering, also called "liberation." An opportunity for liberation presents itself at death. However, the person must recognize the "Clear Light," which may take years of dedicated meditation practice to perfect. One can also attain enlightenment in life, which is recommended. However, according to the Tibetans, only those in precious human bodies have the ability to recognize this Clear Light or "Ground Luminosity" at the instant of death, when the boundless, sky-like nature of mind is uncovered. The best way for a human to prepare for death is to make good use of this life. Aim to live and die without regret, and to practice meditation and become familiar with the true nature of mind — the sky-like background that holds the entire universe in its embrace.

Even though nonhuman animals do not possess the understanding gained in meditation, all living beings, even the smallest insects, go through the same process of dying, even at sudden death. From the Tibetan teachings and from near-death experiences, we find consistent information to help guide a good transition. Importantly, they tell us that, at death, all anger, desire, and ignorance die. Furthermore, the clairvoyant consciousness of dying beings is clearer than in life. They see, hear, and feel those around them, which Lonna also described during her near-death experience when she was kicked by a horse. She saw everything, as if in a mirror. She heard the doctors and nurses and felt their anxiety. The Tibetans explain that the thoughts and emotions of others nearby at the time of death affect the spirit of the dead one.[2] It is important to be harmonious for the peace of mind of the dead one. This is the most important thing I want to stress for the highest good of everyone involved with an animal's death. As traumatic as the event may be, the best thing anyone involved can do is to hold calm, loving thoughts, with prayerful best intentions of forgiveness and gratitude.

The Spirit is eternal. To help comfort and guide our pets, wildlife, and farm animals at death, we must consider the well-being of the Spirit in transition. Hospice is a wonderful option for ailing pets at the end of their lives. For those who die suddenly or traumatically, we still can help. Express gratitude, forgive them and yourself, pray for all karma to be forgiven, and pray they be taken into the Clear Light, the arms of God, or Divine Love. Imagine tremendous rays of light emanating from divine beings, streaming onto the dead one, purifying them, freeing them from confusion and pain. Imagine the dead one healing and dissolving into light.[3] The Tibetans teach that this prayer is beneficial even long after an animal has died.

There are three ways to leave this physical life: to gently fade away, as if falling asleep; to have an assisted death, as in euthanasia; and to experience a traumatic end, such as with disease or an accident. All can be good transitions.

May

I first met May and Jean-Luc on a horseback ride. A mutual friend had invited Jean-Luc, and he rode a pure white mule named May. Jean-Luc and I have been together ever since.

May was a spiritual emanation who blessed our lives; gentle and kind, she carried whoever and whatever we packed on her and followed without being led. Often, she showed us the safest way through or around difficult mountain terrain, a muddy bog, deadfall, snow, or steep slope, as she did on the Quartz Lake pack trip in chapter 9. One of us would look ahead and see May looking back as if to say, "This way." She watched every step she took and carefully placed each hoof. More often than not, we would follow her. May always knew the best way to go.

When she reached thirty years of age, her hind limbs became weak, and she had trouble getting up. Veterinarians know that it sometimes helps to roll a horse over when this happens. One person turns the head at the same time another reaches over from the back of the horse with ropes around the lower parts of the limbs, and together they roll the equine over. Jean-Luc and I rolled

May over many times, and she would stand up and walk off. She learned to roll herself over as well. With age, she became thinner, and Sporty often chased her off the feed. So, Jean-Luc built a barn at the house with individual wooden stalls that had metal hog-wire fencing separating the feed bunks. May could stand right next to Sport and look at him through the hog fencing from her separate stall, nibble grain, and he could not touch her.

One winter's evening, I went out to feed the equines, and May did not show up with the rest of the herd. I hiked through the snow and trees trying to see a white mule in the snow in the dark. I called, "May, May." Then I heard a rustle and found her stuck in a tree well, the back of her neck lay against the trunk of the piñon tree, and her limbs were above her on the snow bank around the tree. I ran to get a halter and yelled to Jean-Luc. He pulled as I pushed, and we slid her on the snow downhill until the momentum got her upright and off she went again.

The next summer she became even thinner. Extra food was no help. I gave her medication for pain every day, and the neighbors started making comments about it being time for her to go. She belonged to Jean-Luc. It was his decision. I mentioned that she was getting really thin. He said nothing. "What do you see when you look at her?" I asked.

"A lot of good memories," he replied.

I let it go and resolved to ignore other people's opinions. This presents a challenge for many people. Friends, neighbors, and relatives all think they know best when someone else should put down their animals, but from what I have seen, they sing a different song when the animal belongs to them. As we learned with old Salty Bones, people do not like to watch an animal suffer, so they want us to please get rid of it. The suffering these humans experience is worse than the animal's in most cases, since it is their judgments that cause them pain. Animals suffer from pain, and that must always be addressed, but they may not always be suffering as much as some observers might think. The animal is just being who it is and living and dying in a natural way. The decision to end an animal friend's life is difficult enough for people to make. No one benefits from the judgmental opinions of others. I have pressured people at times when I perceived the pain to be beyond what medications could abate, but the decision ultimately has to be made by the primary caregiver. I recommend that we not let others pressure us to do something we do not feel good about because, in the end, we each have to live with our decisions.

One day, I drove out to Earl's ranch to give May her medication, check on the other equines, and move the irrigation water. I drove up to the gate and noticed Bonnie, Jean-Luc's big red mule, on the other side of the irrigation ditch, with her head turned to look at me. Her eyes told me that something was wrong. I looked around the ranch and saw May lying behind Earl's house, under a tree, completely still. She had died peacefully, just as she had lived. There was no sign of struggle or injury. May always knew the best way to go. She made a good transition.

Pecos

In veterinary school one day, during freshman year, a fellow student asked a group of us in the coffee room if anyone wanted a horse for cheap. I said, "I do." I bought Pecos for a hundred dollars; he was eighteen months old. He was a bay roan appaloosa who only knew how to rear. His rearing up, with hooves flying in the air, scared me, and it took courage just to go out to his paddock and train him. I could not afford a saddle, and I learned from a well-known horse trainer, Ray Hunt, at a private veterinary school clinic, that I should lean on him and gradually hang over his back, and then climb on top of him, which I did over a period of weeks. Pecos never bucked, and after that, I rode him bareback around the neighborhood. The occasional rear still challenged me. Even so, just sitting on that horse provided a mental break from the stress of school. From his back, I would stare off at the sunset as he wandered around eating.

When I moved to Durango, I finally bought an Australian saddle, and we rode hundreds of miles on day rides and pack trips for about twenty-three years. Pecos loved to run, and he walked faster than most horses. He took on any obstacle I pointed him toward. Pecos was strong, brave, willing, and the most fun I ever had. I remember times that he ran so fast down a sand wash or along a trail that my hat fell down over my eyes, my feet came out of my stirrup, and I had no control. Yet I laughed until I cried because I trusted him so much, and he always kept me safe. He was so much fun!

I finally decided to retire him from riding at age twenty-five. His vision had dwindled and he stumbled sometimes. His stiff joints made the rides a bit rough, and I did not want him to get hurt. I had a mule to ride, so I let him rest. The winter before his thirty-fourth birthday, he injured his left stifle, the joint analogous to the human knee. At that point, I could no longer trim his feet. Because of arthritis in all four limbs, he could not stand on three limbs for a hoof trim. I discovered a helpful tip for anyone with a horse like this. I let him graze freely around the house; he never ran away, and he filed his own feet by walking on the asphalt driveway. He wore them off perfectly.

I treated Pecos with acupuncture, used therapeutic laser, chiropractic, Chinese herbal medicine, and drugs to keep him going. His teeth were worn to the gums. I fed him a fifty-pound bag of senior equine feed each week to keep him from getting too thin. He had a chronic sinus infection with a snotty nose because of bad teeth. The dentist pulled one tooth, but the snot continued and did not improve with medications. He could not see more than a few feet in front of him, and he seemed deaf.

I kept him at the house until midsummer. Just before my annual trip to visit the family in Wisconsin, we took him to Earl's ranch to join the rest of the herd. When he went to drink from the pond, he stumbled and nearly fell in. I had known of too many old horses that drowned in ponds, and that was it for me. Then he walked out onto the bridge, over the irrigation ditch, and called with a weak whinny for the herd. They were right in front of him. Sporty looked at him and nickered, but Pecos did not hear. Sporty looked at me as if to say, "I don't know what to do with him. I can't take care of him anymore." I nodded to Sport in understanding. They had been inseparable pals. In their younger days, they loved to wrestle. I called it "snake dancing." They would rear up together and bite each other in the throats. Pecos seemed to win, but Sport instigated the matches. We had to keep them separated on pack trips or the snake dancing would keep us up all night. I rode Pecos and led Sporty on many pack trips and day rides. In the winter, they snuggled together in the cold, and as Pecos aged, Sporty began to herd him around the ranch and watch after him. In the barn, Sport chased everyone else away from food except for Pecos. Now, he and I agreed. Time was up. We set a day.

Jean-Luc dug a hole with the tractor at the ranch. I fed Pecos all the senior feed he would eat and led him to the ditch for a drink. Then we walked to the grave, and I let him graze on grass along the way. I thanked him over and over for all the wonderful times we shared. I apologized for any times I hurt him. I explained that we could not take care of him anymore, and we had to let him go. I prayed for Divine Love to take him into the arms of bliss and love, and I told Pecos to go to the light. I put on my headphones to protect my ears and told Jean-Luc, "Do it." He paused for a moment with the .357 Magnum placed against Pecos's forehead, then pulled the trigger, putting a bullet in his brain. Pecos liked to move fast, and a bullet was the fastest way to move on.

I walked back to the car because I needed more tissues, and on the way, I felt Pecos walking with me. "Go to the light. Do not stay here. Go home to the Clear Light." Just then Sporty gave a mournful whinny, and Pecos was gone. I prayed for his karma to be purified, and I still pray for him. The Tibetans teach that our prayers, even long after death, help the departed. Visions to use include imagining the loved one healed and being filled with light until they dissolve into pure light. Now I imagine a unicorn's horn on Pecos's forehead, where the bullet entered, and a pure-white, winged Pecos racing off at a gallop, dissolving into Clear Light.

Some insist that we must let animals die on their own even when they suffer the most unpleasant colic or other debility. In this way, we humans can uphold the Buddhist vow to not kill and can avoid bad karma for ourselves, while allowing the animal to suffer its karma in this life and have a better life in the next incarnation. But I offer the following to support assisted death or euthanasia of animals. Each person has to make peace with their own decisions. The guidance for these decisions must come from within. The Buddha said, "You should trust the truth that is within you."[4] Rather than blindly following the dogma or dharma of others, the best thing is to obey our own Buddha-nature, our "inner teacher."[5]

The single most important thing I have learned in my years of veterinary medicine and while writing this book is to trust the guidance of this inner being. Every religion and spiritual teaching refers to it; psychology calls it the "higher conscience," and Jungian psychology calls it the "anima" — the true inner being. Lao-tzu said the Tao is found *within us*.[6] This inner guidance has a voice, clearer than the rest of the mental mumblings. The best way to discern it from the other psychobabble is to pay attention to the voices. This true voice is brutally honest, yet unconditionally loving. The ego cannot trick it with self-pity and blame. It knows everything, sees everything, forgives everything, and warns of consequences along the wrong route. Call it "God," Buddha-nature, higher conscience, or Divine Love, I do not care. Make friends with this voice, learn to trust it, and you will know the correct course of action.

We fear judgment, and we fear the animal will suffer more from our mistakes. Regarding judgment after death, *The Tibetan Book of Living and Dying* and experts on near-death experiences say that the judgments we receive at death come from the individual being judged.[7] We are the judge and the judged. This is why it is so important that we have no regrets. Each individual must decide what is right.

Here is how I judge the transition I chose for Pecos. It was my good karma, and his, that he and I came together. He was my responsibility, and I kept him alive in good condition well beyond what he could have managed without my help. We loved each other and were attached, like one being at times. Holding on to a loved one only hinders their ability to transition. Tibetans tell us to surround the dying with sacred feelings and emotions and to help them let go of yearning, grasping, and attachments.[8] It was Pecos's good karma that I let go of my attachment and had the courage to set him free. It was not easy to say good-bye to Pecos, but I had to cut the ties. Karma is not as important for animals; they are not guilty of sin. No harm comes to the Spirit of any animal because of human action, especially when the beast is loved and surrendered to love. I honor Pecos and am eternally grateful for his friendship and service to me. Pecos was who he was and is destined to be who he is meant to be. Pecos made a good transition.

Sporty

The day I met Sport, I was at a quarter horse breeding ranch with many horses. I walked through the barn looking at all the foals and saw a little gray colt — Sporty. I inquired about him, learned he was for sale, and said, "I'll buy him." Then, I immediately thought, *Who said that?* It was as if some other voice spoke through me. The seller was excited, gave me a five-hundred-dollar discount, and he was mine. Dana, my clairvoyant friend, said Sport put the idea in my head because he knew I would take care of him. And I did. Maybe I owed him, since he needed a lot of care.

Difficult. I have raised a number of horses, and none were anywhere near as difficult as Sporty. Horsemen typically consider all bad horse behavior to be the fault of the person who manages the horse. I struggled with that stigma. However, I was not the only person to find Sport challenging.

A gentle horseshoer trained Sport to halter for a show before I brought him home, and he led well. However, when loose in the pasture, he would chest butt me from behind and bite me. For him it was play, but rough horseplay is dangerous. I smacked him on the lips with a back fist when he bit, and he seemed completely unfazed. Eventually, he decided to just touch me with his mouth and not bite, and I agreed to allow that — he wore me out. Sometimes he kicked at me; his rough behavior could not be tolerated. One day, I set him straight. Looking him in the eye, I said, "Sport, if you ever hurt me, you're dead!" I meant it, and I know he knew it because he quit kicking at me.

While he was still a yearling, Sport would walk into the chest-deep water in the irrigation ditch and go under the fences to visit the neighbor's fillies. I had to castrate him as a yearling to keep him home. I also owned a two-year-old colt at the time, and the two would wrestle all day long, running, rearing, biting, and kicking each other until they were lathered with sweat. The other horses would find a quiet corner out of the way to hide and avoid interacting with the wild maniacs. Sporty loved to fight. He fought with everyone about everything.

The same horseshoer who trained him also had trouble trimming Sport's hooves. One day I was late to a shoeing appointment and found Sporty with the horseshoer's coat over his head and his forelimb bent up in a one-leg hobble. Sporty was bucking in place and squealing. The horseshoer said, "I hope you don't mind; I had to discipline your horse." I understood how difficult Sport could be and said, "I don't mind because you're not hurting him, and he's trying to hurt you." I could not sedate him for every hoof trimming; he had to learn. Usually, if I was present, I could make him behave with a stern word, but sometimes I had to use a twitch or drugs. He gave every new farrier a hard time. Their job is difficult enough, and we need to keep them safe. Eventually, Jean-Luc and I decided to trim Sport's hooves ourselves, and we used rubber boots instead of shoes.

My friend Stacy rode Sport's half brother, and we liked to ride together. That too was dangerous. For example, on one occasion, we were loping along a smooth trail next to a stream, when both horses suddenly spooked, rushed sideways, on opposite sides of the trail, then stopped. Stacy and I looked at each other, shaking our heads, asking, "What was that about?" Fortunately, we both stayed on, but we found it hard to trust those boys.

Sport also suffered numerous health conditions, including a fractured pelvis. When I attempted to treat his mangled pelvis with acupuncture and chiropractic, he swished his tail in my face and tried to step on me. I walked up to his face, placed my index finger on his cheek, and said intensely, "Sport, you have two choices. Get along with me or die. Because I am the only one who gives a spit about you!" He pursed his lips, sighed, and turned his head away. Then he stood still and let me treat him.

After he recovered from the fracture, I was afraid to start riding him again. I had ridden him on several pack trips and many day rides, but he loved to argue, and I did not want another fight. So I hired a gentle cowboy trainer to ride him. When I first introduced the trainer to Sport, he said, "Oh, he's just spoiled." (This fits the typical opinion cowboys have of horses owned by city women.) I told the cowboy that I was afraid he would just get into a fight with Sport. He said, "Ma'am, I don't fight with horses." A week after he had been riding Sport, I called to check on the progress. The cowboy said, "Oh, he's okay; he just doesn't want to do what you want to do sometimes." That was a fact. He continued, "He's not afraid of anything. I had him in a corral surrounded by cattle, and he was calm as could be." Indeed, fear was not Sport's problem. A couple of weeks later, I called again. The cowboy's tone was soft: "Ma'am, I don't normally hit other people's horses, but he just pisses you off enough. I rode him the other day, and my hat came off. I got off to get it, and he wouldn't let me back on, and we went around. Ma'am, I think if you beat that horse every time he did something wrong, it wouldn't do a speck of good. I don't know what to recommend." I brought him home. Sport was an excellent pack horse, even though he liked to bite the mule in front of him in the hocks. Fortunately, Jean-Luc's mule Bonnie never seemed to care.

If you ever wonder why a horse would ride head on into war, to face battle among swords, guns, and other horses, know that there are beasts like Sport. He was a war horse. He was always ready to take on a fight.

Sport's many health issues were just as challenging to manage as Sporty was himself, such as dry eye, which is rare in horses. When I finally figured out that dry eye was causing his swollen, red, glop-filled eyes, I learned that I had to treat them with Sport's own serum. I collected blood from his jugular vein, spun it down to separate the red cells from the serum, and drew up the clear, gold liquid into 1-cc syringes. I stored a number of syringes in the refrigerator, and then used one to squirt a half cc of serum into each eye. At first, I did this procedure multiple times a day, then it was once daily for years. Although the serum did not hurt him (it had to feel good), and it helped tremendously, Sport made treatment difficult. I had to snub his head tight to a strong cross post, and he still made application challenging by squinting his eyes and turning his head side to side. Serum would fly onto his face, missing the eye, and I would scream at him, "I'm trying to help you!" Eventually, after years of treating him, I was able to walk out to the pasture, remove his fly mask, put on a halter, and squirt the serum into each eye without a fight.

A number of my friends told me they would have been rid of Sport years ago, but I still loved him and he taught me a lot. I became quite the expert on palpating horses for broken pelvises.

A couple of summers after Pecos died, I went to Earl's to treat Sport's eyes as usual and found him standing on the other side of the bridge, covered in mud. His facemask was caked solid with clay, and he stood absolutely still. I started laughing. "What did you do to yourself this time?" I took the mask off and saw that his eyes were swollen nearly shut, and he looked depressed; his gums were a bit tan. The only sign of trauma was a small laceration on the inside of his right hindlimb, above the ankle. I tugged at a piece of skin on his neck — the skin did not recoil but stayed up in the pinch. He was severely dehydrated and in shock. He walked to the barn without trouble, but he started to look at his sides, first one side, then the other. After brushing off the mud to look for injuries, I saw no sign of trauma. I passed a stomach tube and pumped three gallons of water in him just as Jean-Luc arrived. Then Sporty stretched out and urinated a long stream of blood. His heart rate was only mildly elevated, and he did not paw or roll. He just stood there, uninterested in food. We decided to take him home so I could treat him. When he balked at loading in the trailer, I said, "I promise, I will bring you back to the ranch." Then he loaded right away.

I gave him intravenous fluids and anti-inflammatories, performed acupuncture, and did rectal exams. His colon was displaced, and we took another trailer ride to try to shake it back in place. He never acted painful like an intestinal colic, but he continued to urinate blood. Kidney damage from shock may have been the cause. He made urine, and he passed a full bladder's worth of fluid. Yet with no signs of major trauma, I saw no way to help; he was dying. My friend Jan, a fellow veterinarian in a city nearby, consulted with me, and we concluded that hospitalization and/or exploratory surgery seemed destined to end in death. Sport was twenty-two years old, with eyes that hurt him all the time. I saw no point in attempting heroics. Back at the barn, I encouraged Sport to rally and heal. I looked into the slit between Sporty's eyelids and heard him say, *Let me go. It's time. I decided.* We both knew this problem was not fixable. He always wanted to do things his way, and since he was calm, I let him go into the barnyard to make his own natural transition.

At 2 AM, I went out to give him more pain relief and found him drinking from the water tank and nibbling on hay, but he was not present in his body. He did not seem to notice me, and his body felt ice cold. I gave the injection, and through my tears, I thanked him for being such an entertaining friend and great teacher. I apologized for anything I did that hurt him. I prayed that his karma be forgiven and that the karma between us be forgiven, and I told him to go to the light. At 5 AM, he was dead. He never struggled with the environment, just laid down and let go. Sporty finally had to accept surrender, and he appeared to have found bliss because the expression on his face displayed complete euphoria. I had never seen that facial expression on him before, or on any other dead horse, for that matter. Wherever he had gone, it was a happy place.

Jean-Luc planned to bury Sport at the house, but I shook my head and said, "I promised I'd take him back to the ranch." It took most of the day, but Jean-Luc moved Sport's body and buried him next to his pal, Pecos.

Sport died like a good Buddhist, suffering his karma. I suspect that he and I had an agreement to work together in this life. Hopefully, he has an easier time in the next. He owes me nothing. He went out the way he lived, and for him, it was a good transition.

Since animals are not responsible for sin, and because the idea of "good" belongs to humans, all animal transitions happen as they should, sometimes in ways and for reasons we may not understand. We must be grateful for the service and friendship animals provide, and celebrate their lives, honor loving memories, and have no regrets. To dissolve regrets, offer prayers for the departed. All animal spirits return to Source Spirit. All bodies eventually return to the womb of Mother Earth. All organic matter degrades into nutrients to feed seeds as they burst open in rebirth. All are recycled and reborn at the right time, in the right way. All will be who they are — beautiful expressions of a combination of the Holy Spirit and Mother Nature. All is well.

Notes

Chapter 1. Ambulatory Horse Doctor

1. According to the *Utah Historical Quarterly*, in 1765, the Spanish explorer Juan Maria Antonia de Rivera named the river River de Las Animas, meaning "gives force or spirit," because it was very swift and difficult to cross. The common misnomer Rio de Las Animas Perdidas, or River of Lost Souls, came later or more correctly fit Purgatory Creek. See *Utah Historical Quarterly* 60, no. 3 (summer 1992): 2008.
2. Anne Bancroft, ed., *The Buddha Speaks* (Boston: Shambhala Publications, 2000), 83.
3. Thomas Cleary, trans., *Living a Good Life* (Boston: Shambhala Publications, 1997), 32.
4. Bancroft, *Buddha Speaks*, 99.
5. Eckhart Tolle, *The Power of Now* (Novato, CA: New World Library, 1999), 128.

Chapter 2. The Beginning: Creation and the Garden Paradise

1. Mircea Eliade, *Shamanism: Archaic Techniques of Ecstasy* (Princeton, NJ: Princeton University Press, 1974), 99.
2. Donald B. Redford, ed., *The Ancient Gods Speak* (Oxford: Oxford University Press, 2002), 239.
3. Richard Heinberg, *Memories and Visions of Paradise* (Los Angeles: Jeremy P. Tarcher, 1989), 15–166.
4. L. Robert Keck, *Sacred Quest: The Evolution and Future of the Human Soul* (West Chester, PA: Chrysalis Books, 2000), 54.
5. Frank Waters, *Book of the Hopi* (New York: Penguin Books, 1977), 1–22. Some scholars find this information unreliable, although Waters states that it was received from full-blooded Hopi people living on the reservation and that his translator was

also a full-blooded Hopi born in Oraibi. Via email (October 11 and 16, 2001), I questioned Ekkehart Malotki of Northern Arizona University, who is an expert on Hopi language. His reply is pertinent to any religious story: "I would assume that sections relating to oral literature (myths, stories, legends) would be okay. However, you must keep in mind that there is not just one version of a belief, mythological event, etc. Cultures are living organisms; therefore, they are in constant flux. One storyteller's memory may be more accurate than that of another, or he may simply be a better storyteller.... Oral stories vary from clan to clan and from village to village. I agree with you that there is no agreement on anything in religion, or anything else."

6. Waters, *Book of the Hopi*, 12.

7. According to Ekkehart Malotki, the correct spelling is Lalvayhoya, which means "talkative person." Mochni is correctly spelled *motsni* and is described in the Hopi dictionary as a "loggerhead shrike." A mockingbird is *yappa*, a bird that can mimic all other birds. The Hopi themselves are considered to be like mockingbirds in that they frequently mimic the songs and sounds of other peoples.

8. Malotki states that the correct spelling for this snake is *Ka'toya*, which is a monstrous snake. It is not a Hopi word but a word borrowed from another Native American language.

9. From "Noah's Ark on Ararat," *Encounters with the Unexplained*, directed by David Priest, aired May 25, 1999 (Baker, OR: Grizzly Adams Productions, 2001), DVD. This video was based in part on the book by Charles E. Sellier and David W. Balsiger, *The Incredible Discovery of Noah's Ark* (New York: Dell, 1995).

10. Robert M. Hazen, "The Joy of Science," lecture 57, *Great Courses Magazine* (October 2006), 13.

11. Robert Lawlor, *Voices of the First Day: Awakening in the Aboriginal Dreamtime* (Rochester, VT: Inner Traditions International, 1991).

12. Lawlor, *Voices*, 36. Lawlor quotes Nancy C. Munn, "The Transformation of Subjects into Objects in Walbiri and Pitjandtjartjara Myths," 62. The "primal note" of creation is also mentioned by Caitlin Matthews in *The Encyclopedia of Celtic Wisdom* (Rockport, MA: Element, 1994), 221. The word *enchant* literally means "to infuse with song." As Matthews writes, "Music is the first ordering of chaos."

13. Lawlor, *Voices*, 15, 17.

14. Ibid., 115. The author makes an interesting comparison between this serpent myth and the serpent symbolism of ancient Egypt.

15. K. Langloh Parker, *Australian Legendary Tales* (Sydney: Angus & Robertson, 1974), 9–10.

16. Lawlor, *Voices*, 1, 69.

17. Jeremy Narby, *The Cosmic Serpent, DNA, and the Origins of Knowledge* (New York: Jeremy P. Tarcher/Putnam, 1999), chapter 6.

18. David R. Hawkins, *Transcending the Levels of Consciousness: The Stairway to Enlight-enment* (West Sedona, AZ: Veritas, 2006), 33.

19. Heinberg, *Memories and Visions*, 88, 89, 90.

20. Ibid., 91.

21. Elena Marsella, *The Quest for Eden*, as quoted in Heinberg, *Memories and Visions*, 92.

22. Heinberg, *Memories and Visions*, 82–84.

23. Maria Leach, *The Beginning* (New York: Funk & Wagnalls, 1956), 143–44, as quoted in Heinberg, *Memories and Visions*, 84.

24. Keck, *Sacred Quest*, 38–89, 45.

25. Joseph Epes Brown, *The Sacred Pipe* (Norman, OK: University of Oklahoma Press, 1989), 4n.

Chapter 3. The Hunter-Gatherer: Shamanism

1. Quote is from the Pawnee tribe, as transmitted to Natalie Curtis, c. 1904, in Joseph Campbell, *The Way of the Animal Powers* (San Francisco: Harper & Row, 1983), 1, 8.

2. Mircea Eliade, *Shamanism: Archaic Techniques of Ecstasy*, trans. Willard R. Trask (Princeton, NJ: Princeton University Press, 1964), 486.

3. Eliade, *Shamanism*, 508–9.

4. Jean Clottes and David Lewis-Williams, *The Shamans of Prehistory: Trance and Magic in the Painted Caves*, trans. Sophie Hawkes (New York: Harry N. Abrams, 1998), 114.

5. Eliade, *Shamanism*, 508.

6. Ibid., 478.

7. Ibid., 479.

8. Brown, *Sacred Pipe*, 39ff. (see chap. 2, n. 25).

9. Research of the neuropsychological condition of trance is explained in Clottes and Lewis-Williams, *Shamans of Prehistory*, 16–17.

10. Narby, *Cosmic Serpent*, 24 (see chap. 2, n. 17).

11. John G. Neihardt, *Black Elk Speaks* (Lincoln: University of Nebraska Press, 1961), 23–24.

12. Clottes and Lewis-Williams, *Shamans of Prehistory*, 17.

13. Ted Andrews, *The Art of Shapeshifting* (Jackson, TN: Dragonhawk Publishing, 2005), 224.

14. Brown, *Sacred Pipe*, 5ff.: "*Wakan-Tanka* as Grandfather is the Great Spirit independent of manifestation, unqualified, unlimited, identical to the Christian Godhead, or to the Hindu *Brahma-Nirguna*. *Wakan-Tanka* as Father is the Great Spirit considered in relation to His manifestation, either as Creator, Preserver, or Destroyer, identical to the Christian God, or the Hindu *Brahma-Saguna*."

15. J. O. Dorsey, "A Study of Siouan Cults," in Brown, *Sacred Pipe*, 6.

16. Brown, *Sacred Pipe*, 31.

17. In numerous other cultures, eight-legged beasts are shamanic. For example, Odin of ancient Germany as well as Siberian shamans rode eight-footed horses into the spirit realm.
18. Brown, *Sacred Pipe*, 85.
19. Ibid., 52.
20. Ibid., 62.
21. Eliade, *Shamanism*, 184.
22. Ibid., 69.
23. Ibid., 185.
24. Ibid., 174.
25. Narby, *Cosmic Serpent*, 152.
26. Konrad Zacharias Lorenz, an Austrian zoologist, ethologist, and ornithologist, received the Nobel Prize in Physiology or Medicine in 1973. According to animal behavior scientist Marc Bekoff, Lorenz believed animals experience many of the same emotions as humans. The body map is one's self-representation in one's own brain, one's assumption or conception of what one's body is like. See Marc Bekoff and Colin Allen, *Species of Mind* (Cambridge, MA: Bradford Book, MIT Press, 1999), 30–31.
27. Calvin Martin, *Keepers of the Game: Indian-Animal Relationships and the Fur Trade* (Berkeley: University of California Press, 1978), 20–21n.
28. Dr. Donald Noah, "Bioterrorism Still a Threat, but Veterinarians Can Help," *Journal of the American Veterinary Medical Association* 219, no. 6 (September 15, 2001): 717.
29. Martin, *Keepers of the Game*, 131–32.

Chapter 4. Mother Nature: The Great Transformer

1. W. R. Halliday, *The Pagan Background of Early Christianity* (New York: Cooper Square Publications, 1970), 110. Halliday was a historian and archeologist.
2. Ken Dowden, *European Paganism* (New York: Routledge, 2000), 15.
3. Mircea Eliade, *A History of Religious Ideas*, vol. 1, trans. Willard R. Trask (Chicago: University of Chicago Press, 1978), 4.
4. Ibid., 5.
5. Halliday, *Pagan Background*, 168–72.
6. Walter Burkett, *Greek Religion* (Cambridge: Harvard University Press, 1985), 59.
7. John Bowker, ed., *The Oxford Dictionary of World Religions* (Oxford: Oxford University Press, 1997), 569, 834. John the Baptist calls Jesus "the lamb of God" (John 1:29–36), and in the book of Revelation, the lamb appears repeatedly as Christ.
8. Neil Gordon Munro, *The Ainu Bear Ceremony* (Great Britain and Ireland: Royal Anthropological Institute, 1931), DVD.
9. Joseph Campbell, *The Way of the Animal Powers* (San Francisco: Harper & Row, 1983), 1:154.

10. Seanicaa Edwards and Ray Massey, "The Impact of Livestock Production on Local Economics: Summary of Literature," University of Missouri Extension, posted October 2011, accessed March 23 to April 10, 2017, http://extension.missouri.edu /p/MP752.

11. Kim Dillivan and Jack Davis, "Economic Benefits of the Livestock Industry," Igrow .org, posted July 7, 2014, accessed March 23 to April 10, 2017, http://igrow.org /livestock/profit-tips/economic-benefits-of-the-livestock-industry/7/7/2014.

12. AVMA Animal Welfare Principles, accessed March 23 to April 10, 2017, https://www.avma.org/KB/Policies/Pages/AVMA-Animal-Welfare-Principles.aspx.

13. BSEinfo, accessed March 23 to April 10, 2017, http://www.bseinfo.org/beefindustry facts.aspx.

14. US Department of Agriculture, "Cattle: Background," posted January 23, 2017, accessed March 23 to April 10, 2017, www.ers.usda.gov/topics/animal-products /cattle-beef/background.aspx. As disturbing as a feedlot might appear to some, it may be a reflection of American society, which is concentrated in crowded cities and in which 35.7 percent of adults are obese.

15. Catherine Green, "Organic Market Overview," USDA Economic Research Service, last updated April 4, 2017, www.ers.usda.gov/topics/natural-resources-environment /organic-agriculture/organic-market-overview.

16. Nancy Gagliardi, "Consumers Want Healthy Foods — and Will Pay More for Them," Forbes (February 18, 2015), https://www.forbes.com/sites/nancygagliardi /2015/02/18/consumers-want-healthy-foods-and-will-pay-more-for-them/#68f8 011475c5.

17. Marija Gimbutas, *The Living Goddesses*, ed. Miriam R. Dexter (Berkeley: University of California Press, 2001), 5.

18. Laurence Gardner, *Genesis of the Grail Kings* (Boston: Element Books, 2000), 203. Gardner writes, "Those who communicated directly with the gods were generally attributed with horns; it was for this reason that Michelangelo (1475–1564) added horns to his famous statue of Moses on the Roman monument of Pope Julius II." Another example is the Hopi One and Two Horn priests.

19. Stephen E. Thompson quoted in *The Ancient Gods Speak: The Guide to Egyptian Religion*, ed. Donald B. Redford (Oxford: Oxford University Press, 2002), 70.

20. Bob Brier, "Animal Mummies," lecture 45, *The History of Ancient Egypt* (Chantilly, VA: The Teaching Company, 1999), DVD.

21. Ibid.

22. Roel Sterckx, *The Animal and the Daemon in Early China* (Albany: State University of New York Press, 2002), 5.

23. Richard Wilhelm writes, "Its origin goes back to mythical antiquity, and it has occupied the attention of eminent scholars of China down to the present day. Nearly all that is greatest and most significant in the three thousand years of Chinese cultural

history has either taken its inspiration from this book, or has exerted an influence on the interpretation of its text. Therefore it may safely be said that the seasoned wisdom of thousands of years has gone into the making of the *I Ching*. Small wonder then that both of the two branches of Chinese philosophy, Confucianism and Taoism, have their common roots here." Richard Wilhelm, *The I Ching* (New York: Bollingen Foundation, 1950), xlvii.

24. Martin Schönberger, *The I Ching and the Genetic Code: The Hidden Key to Life* (Santa Fe, NM: Aurora Press, 1992), 147.

25. Sterckx, *Animal and the Daemon*, 100.

26. Ibid., 99.

27. Ibid., 168; *Zhuangzi jishi*, 18.624–25.

28. Ibid., 81.

29. Ibid., 77.

30. *The Yellow Emperor's Classic of Medicine*, trans. Maoshing Ni (Boston: Shambhala, 1995), 19.

31. Giovanni Maciocia, *The Foundations of Chinese Medicine* (New York: Churchill Livingstone, 1989), 73.

32. Sterckx, *Animal and the Daemon*, 130.

33. E. A. Wallis Budge, *The Gods of the Egyptians*, vol. 1 (New York: Dover Publications, 1969), 11, 24.

34. Josh Gabbatiss, "The Lost Giants that Prowled the Australian Wilderness," *BBC*, February, 8, 2016, http://www.bbc.com/earth/story/20160208-the-lost-giants-that-prowled-the-australian-wilderness.

35. Lawlor, *Voices*, 30 (chap. 2, n. 11).

36. Sterckx, 180; *Ersanzi wen*, 424, lines 6–7; *Boshu Zhouyi jiaoshi*, 348–49.

37. Karen Armstrong, *A History of God* (New York: Ballantine Books, 1993), 10.

38. Gardner, *Genesis of the Grail Kings*, 35.

39. Sterckx, *Animal and the Daemon*, 180.

40. In the nineteenth century a German biologist, Ernst Haeckel, noted embryological parallelism, which lead to the theory that "ontogeny recapitulates phylogeny," meaning that the developmental phases of an embryo follow the same forms as seen in the evolution of species. Haeckel overemphasized the similarities in his drawings, and this notion was disproved. However, in general, embryos are difficult to tell apart. I showed my embryology teacher eight drawings of embryos in the tail-bud stage and asked him if he could identify them. He was correct about the general classes — fish, reptile, and mammal — of several, and he agreed that the average biologist would only be able to go so far as to identify the group as embryos. All start from two cells and develop into stages with a yolk sac, gill slits, and a tail. All embryos, at one point, resemble tadpoles.

41. Quote from the Peshitta Bible, which is similar to the King James Version.

42. Sterckx, *Animal and the Daemon*, 178.

43. Ibid., 201; *Lunheng jiaoshi*, 50.732–33.

44. Ibid., 86; *Da Dai Liji*, 5.8b–9a.

45. Ibid., 84; *Huainanzi*, 4.154–55.

46. Ibid., 208; *Zuozhuan zhu*, 197; *Qianfu lun*, 6.358.

47. Ibid.

48. Ibid., 208; *Lunheng jiaoshi*, 64.925.

49. Ibid., 237.

50. Ibid., 154; *Kongzi jiayu*, 5.9b.

51. *Yellow Emperor's Classic of Medicine*, 1.

52. Sterckx, *Animal and the Daemon*, 212–13; *Huainanzi*, 13.457-59.

53. Ibid., 26.

54. Dowden, *European Paganism*, 156–59.

Chapter 5. Hinduism: Escape from Samsara

1. Deepak Chopra, *How to Know God: The Soul's Journey into the Mystery of Mysteries* (New York: Three Rivers Press, 2000), 51.

2. Mircea Eliade, *A History of Religious Ideas*, trans. Willard R. Trask (Chicago: University of Chicago Press, 1978), 1:196.

3. The sayings of Shri Ramakrishna quoted in Merle Severy, ed., *The World's Great Religions* (New York: Time Incorporated, 1957), 38.

4. Trilok Chandra Majupuria, *The Sacred Animals of Nepal and India* (Bangkok: Craftsman Press, 2000), 19.

5. Amiya Chakravarty, "The Quest for the Universal One," in *Great Religions of the World* (Washington, DC: National Geographic Society, 1971), 34.

6. From the Chandogya Upanishad, quoted in Surendranath Dasgupta, *A History of Indian Philosophy* (Cambridge, UK: Cambridge University Press, 1969), 38.

7. Sarvepalli Radhakrishnan, trans., *The Brahma Sutra* (New York: Harper & Brothers, 1960), 205–6.

8. *Bhagavad-Gita As It Is*, ed. A. C. Bhaktivedanta Swami Prabhupada (Los Angeles: The Bhaktivedanta Book Trust, 1972), 223.

9. Ibid., 171.

10. Ibid., xxvi.

11. Ibid., 57.

12. Ibid., xvi, 32.

13. Ibid., 235.

14. Sir Charles Eliot, *Hinduism and Buddhism: An Historical Sketch*, vol. 3 (London: Routledge & Kegan Paul, 1921), 42n, found in Jataka 159.

15. For instance, see the following: F. Buch, et al., "The Quantification of Bone Tissue

Regeneration after Electromagnetic Stimulation," *Archives of Orthopaedic and Trauma Surgery* 112, no. 2 (1993): 75–78; L. P. Chen, Z. B. Han, and X. Z. Yang, "The Effects of Frequency of Mechanical Vibration on Experimental Fracture Healing," *Zhonghua Wai Ke Za Zhi (Chinese Journal of Surgery)* 32, no. 4 (1994): 217–19; T. F. Cook, "The Relief of Dyspnea in Cats Purring," *New Zealand Veterinary Journal* 21 (1973): 53–54; L. M. Cristiano and R. M. Schwartzstein, "Effect of Chest Wall Vibration on Dyspnea during Hypercapnia and Exercise in Chronic Obstructive Pulmonary Disease," *American Journal of Respiratory Critical Care Medicine* 155, no. 5 (1997): 1552–59; and M. Falempin and S. F. In-Albon, "Influence of Brief Daily Tendon Vibration on Rat Soleus Muscle in Non-Weight-Bearing Situation," *Journal of Applied Physiology* 87, no. 1 (1999): 3–9.

16. Melvin Morse quoted in Betty J. Eadie, *Embraced by the Light* (New York: Bantam Books, 1994), xiv–xv.

17. Majupuria, *Sacred Animals*, 73, 108.

18. Thomas Ashley-Farrand, "Seed Mantras," session 2, *Mantra: Sacred Words of Power* (Boulder: Sounds True, 1999), audiocassette.

19. Eadie, *Embraced*, 45.

20. Severy, *World's Great Religions*, "The Truth behind the Veil: Hindus Use Myths, Symbols and an Elaborate Theology to Explain Many Sides of Oneness," 16.

21. Ashley-Farrand, "Seed Mantras."

22. Majupuria, *Sacred Animals*, 133.

23. Ibid., 79.

24. Poulomi Banerjee, "Hinduism vs. Hindutva: The Search for an Ideology in Times of Cow Politics," *Hindustan Times*, April 10, 2017, www.hindustantimes.com /india-news/hinduism-versus-hindutva/story-SYB9a5bwKPqBJxbM4fPg2O.html.

25. Sadhguru, "Can You Enlighten an Animal?" parts 1 and 2 (Isha Foundation), YouTube, posted May 18, 2010, https://www.youtube.com/watch?v=G076TLneqro.

26. Prabhupada, *Bhagavad-Gita*, 90.

27. *Great Religions of the World*, 164.

28. Prabhupada, *Bhagavad-Gita*, 216.

29. Quote is cited in Paramahansa Yogananda, *Autobiography of a Yogi*, 11th ed. (Los Angeles: Self-Realization Fellowship, 1974), 187n.

30. Deepak Chopra, session 4, *Magical Mind, Magical Body* (Niles, IL: Nightingale Conant, 1990), audiocassette.

31. Michael W. Fox, letter to the editor, *Journal of the American Veterinary Medical Association*, May 2002.

32. Allen M. Schoen, *Kindred Spirits* (New York: Broadway Books, 2001). I recommend the entire book for anyone wanting more about the history of how veterinary medicine has changed with regard to treating pain.

33. Mary Lutyens, *Krishnamurti: The Years of Fulfillment* (New York: Farrar, Straus & Giroux, 1983), 62.

Chapter 6. Buddhism: Finding Peace of Mind

1. John Bowker, ed., *The Oxford Dictionary of World Religions* (Oxford: Oxford University Press, 1997), 168.
2. Thich Nhat Hanh, *The Heart of the Buddha's Teaching* (New York: Broadway Books, 1999), 6–7.
3. Fritjof Capra, *The Tao of Physics* (Boston: Shambhala, 2000), 80, 131. Noted physicists Niels Bohr and Julius Robert Oppenheimer are quoted to say that atomic physics was already understood by the Far Eastern religions of Hinduism, Buddhism, and Taoism, among others.
4. The Dalai Lama, *A Simple Path* (London: Thorsons, 2000), 17.
5. Sogyal Rinpoche, *The Tibetan Book of Living and Dying* (San Francisco: HarperSanFrancisco, 1993), 46.
6. Fritjof Capra, *Tao of Physics: An Exploration of the Parallels between Modern Physics and Eastern Mysticism* (Boston: Shambhala, 2000), 24.
7. Sogyal Rinpoche, *Tibetan Book*, 47.
8. Thich Nhat Hahn, *The Art of Mindful Living* (Boulder, CO: Sounds True, 1991), part 1, side A, audiocassette.
9. Hanh, *Heart of the Buddha's Teaching*, 6–7.
10. Lama Surya Das, *Awakening the Buddha Within* (New York: Broadway Books, 1997), 139.
11. Ananda K. Coomaraswamy, *Buddha and the Gospel of Buddhism* (New York: Harper & Row, 1916), 288–89.
12. Noor Inayat Khan, ed., *Twenty Jataka Tales* (Rochester, VT: Inner Traditions International, 1975), 15–21.
13. Coomaraswamy, *Buddha and the Gospel*, 13, 14.
14. Ibid., 11. Coomaraswamy explains that this circumstantial biography of the Buddha is an expression of the facts understood by followers of Gautama. He takes his material from various sources, chiefly the *Nidānakathā*, the *Mahā Parinibbāna Sutta*, and the *Lalitavistara*.
15. Khenpo Karthar Rinpoche, *Dharma Paths* (Ithaca, NY: Snow Lion Publications, 1992), 49.
16. Majupuria, *Sacred Animals*, 101.
17. Hanh, *Heart of the Buddha's Teaching*, 3.
18. Bancroft, *Buddha Speaks*, 91, 111.
19. Eliot, *Hinduism*, 1:xxi.
20. Coomaraswamy, *Buddha and the Gospel*, 214.

21. Peter Harvey, *An Introduction to Buddhism, Teaching History, and Practices* (Cambridge, UK: Cambridge University Press, 1990), 32.

22. The Dalai Lama, *Simple Path*, 47.

23. Sogyal Rinpoche, *Tibetan Book*, 112.

24. Harvey, *Introduction to Buddhism*, 33–34.

25. Ibid., 41.

26. Malcolm David Eckel, "The Path to Nirvana," lecture 6, *Buddhism* (Chantilly, VA: The Teaching Company, 2001), audiocassette.

27. The Dalai Lama, *Simple Path*, 40.

28. For one description of the causes of suffering, see Malcolm David Eckel, "All Is Suffering," lecture 5, *Buddhism* (Chantilly, VA: The Teaching Company, 2001), audiocassette.

29. Hanh, *Heart of the Buddha's Teaching*, 27.

30. Bancroft, *Buddha Speaks*, 14.

31. Hanh, *Heart of the Buddha's Teaching*, 42.

32. Bancroft, *Buddha Speaks*, 93.

33. Harvey, *Introduction to Buddhism*, 202.

34. Ibid., 203.

35. Ibid., 204.

36. Christopher Chapple, "Inherent Value without Nostalgia: Animals and the Jaina Tradition," in Paul Waldau and Kimberly Patton, eds., *A Communion of Subjects: Animals in Religion, Science and Ethics* (New York: Columbia University Press, 2006), 248.

37. A Tibetan Buddhist teacher passed this information along to me in a class on *The Tibetan Book of Living and Dying* by Sogyal Rinpoche, citing *The Treasury of Precious Qualities* by Longchen Yeshe Dorje.

38. *Great Religions of the World*, 164 (see chap. 5, n. 5).

39. Tsultrim Allione, "Tonglen Durango" (lecture, the Smiley Building, Durango, CO, March 20, 2004). Tsultrim was awarded the title of "Lama" in 2008.

40. L. Austine Waddell, *The Buddhism of Tibet, or Lamaism* (Cambridge: W. Heffer & Sons Limited, 1967), 125. The Avatansaka Sutra by Nagarjuna is from Samuel Beal, *A Catena of Buddhist Scriptures from the Chinese* (London: Trubner, 1871), 125.

41. Allione, "Tonglen Durango," March 2004.

42. Das, *Awakening the Buddha*, 18.

43. Stories of Padmasambhava are taken from Eckel, "The 'First Diffusion of the Dharma,' in Tibet," lecture 16, *Buddhism*, and from stories told by Lama Lhanang on retreat at Blue Lake Ranch, Hesperus, CO, June 2004.

44. Tsultrim Allione, "Enlightenment through Embodiment" and "Exploring the Dakini Principle," *The Mandala of the Enlightened Feminine* (Boulder, CO: Sounds True, 2003), CD.

45. Eckel, "Buddhist Tantra," lecture 14, *Buddhism*.

46. Waddell, *Buddhism of Tibet*, 31.

47. Lama Lhanang, on retreat, June 2004, and Sogyal Rinpoche, *Tibetan Book*, 191–92.

48. Khenpo Karthar Rinpoche, *Dharma Paths*, 70.

49. Das, *Awakening the Buddha*, 202.

50. Harvey, *Introduction to Buddhism*, 279; Eckel, "Zen," lecture 23, *Buddhism*.

51. Eckel, "The Classical Period of Chinese Buddhism," lecture 20, *Buddhism*.

52. Ibid.

53. Jean Smith, *The Beginner's Guide to Zen Buddhism* (New York: Bell Tower, 2000), 132. This actually comes from the Heart Sutra. Also, see Gichin Funakoshi, *Karate-Dō Kyōhan*, trans. Tsutomu Ohshima (Tokyo: Kodansha International, 1990), 4.

54. Funakoshi, *Karate-Dō Kyōhan*.

55. Coomaraswamy, *Buddha and the Gospel*, 254–55.

56. Ibid., 171, 172.

57. Jakusho Kwong, "Turning Your Light Inward," session 11, *Breath Sweeps Mind* (Boulder, CO: Sounds True, 2003), audiocassette.

58. Hanh, *Heart of the Buddha's Teaching*, 159.

59. Shunryu Suzuki-roshi, session 3, *Zen Mind Beginner's Mind* (Audi Audio/Phoenix Books, 1988), audiocassette.

60. Smith, *Beginner's Guide*, 148.

61. Thich Nhat Hanh, *The Art of Mindful Living* (Boulder, CO: Sounds True, 1991), audiocassette.

62. Smith, *Beginner's Guide*, 116–17.

Chapter 7. Islam, Judaism, and Christianity: The God of Abraham

1. Gershon Winkler, *Magic of the Ordinary: Recovering the Shamanic in Judaism* (Berkeley, CA: North Atlantic Books, 2003), 35.

2. Alan Cooperman and Gregory A. Smith, "How Many People Would Say That They Believed in God If They Were Able to Answer with Complete Anonymity?" Pew Research, December 21, 2010, http://www.pewresearch.org/2010/12/21/how-many-people-would-say-that-they-believed-in-god-if-they-were-able-to-answer-with-complete-anonymity.

3. Pew Research Center, "The Global Religious Landscape," December 18, 2012, http://www.pewforum.org/2012/12/18/global-religious-landscape-exec; this survey reported 31.5 percent Christians, 23.2 percent Muslims, and 0.2 percent Jews. Conrad Hackett and David McClendon, "Christians Remain World's Largest Religious Group, But Are Declining in Europe," Pew Research Center, April 5, 2017, http://www.pewresearch.org/fact-tank/2017/04/05/christians-remain-worlds-largest-religious-group-but-they-are-declining-in-europe; this survey reported 2.3 billion Christians, 1.8 billion Muslims, and 10 million Jews in the world.

4. Bowker, *Oxford Dictionary*, 10–11.
5. Bart D. Ehrman, "Do We Have the Original New Testament?," lecture 24, *The New Testament* (Chantilly, VA: The Teaching Company, 2000), audiocassette.
6. Bart D. Ehrman, "The Copyists Who Gave Us Scriptures," lecture 9, *The History of the Bible: The Making of the New Testament Canon* (Chantilly, VA: The Teaching Company, 2005), audiocassette.
7. Bowker, *Oxford Dictionary*, 888–89.
8. Phillip Cary, "Medieval Christian Theology: Nature and Grace," lecture 14, *Philosophy and Religion in the West* (Chantilly, VA: The Teaching Company, 1999), audiocassette.
9. Bart D. Ehrman, "Lost Christianities: Christian Scriptures and the Battles over Authentication," *Great Courses Magazine* (Chantilly, VA: The Teaching Company, November 2006), 8.
10. Phillip Cary, "Platonist Philosophy and Scriptural Religion," lecture 7, *Philosophy and Religion in the West* (Chantilly, VA: The Teaching Company, 1999), audiocassette.
11. Luke Timothy Johnson, "The Resurrection Experience," lecture 12, *Early Christianity: The Experience of the Divine* (Chantilly, VA: The Teaching Company, 2000), audiocassette.
12. Reverend Ed Beck, email to author, May 19, 2005.
13. Winkler, *Magic of the Ordinary*, 17–18; quotes from Rabbi Moshe Cordovero, *Shi'ur HaKomah*, *Torah*, and Sefer HaZohar, *Midrash HaNe'elam*.
14. Bowker, *Oxford Dictionary*, 71.
15. Amy-Jill Levine, "Adam and Eve," lecture 2, *The Old Testament* (Chantilly, VA: The Teaching Company, 2000), audiocassette and course guidebook, 7.
16. Ibid., 10.
17. Winkler, *Magic of the Ordinary*, 173–74; Moshe Cordovero, *Shi'ur Komah*.
18. Abdullah Yusuf Ali, *The Qur'an: A Guide and Mercy* (Elmhurst, NY: Tahrike Tarsile Qur'an, 2014); regarding forbidden food: 2:173; 5:3; 16:115; regarding use of animals for food and riding: 16:5, 8, 14; 22:28, 34; 23:21; 35:12.
19. Rabbi Roller quoted on now-defunct website www.askarabbi.com, posted May 19, 2003, accessed September 23, 2005.
20. Winkler, *Magic of the Ordinary*, 167.
21. David A. Cooper, "*Techiat HaMaytim:* God Consciousness," session 12, *The Holy Chariot: Practices on the Jewish Mystical Path to Higher Consciousness* (Boulder, CO: Sounds True, 1997), audiocassette.
22. Winkler, *Magic of the Ordinary*, 170; Rabbi Yehudah Loew in *B'er HaGolah*.
23. L. J. Keelin, L. Jonare, and L. Lanneborn, "Investigating Horse-Human Interactions: The Effect of a Nervous Human," *Veterinary Journal* 181, no. 1 (July 2009): https://www.ncbi.nlm.nih.gov/pubmed/19394879. The handler was told to lead the

horse down an alley to a point where an umbrella would be opened. The umbrella never opened, but the human anticipated the horse reacting to it and had an increased heart rate, which the horse sensed.

24. Dr. Kit Flowers, "Listening to Balaam's Donkey," *Christian Veterinarian: The Journal of Christian Veterinary Mission* (April 2001), 2.

25. Jay McDaniel, "Practicing the Presence of God: A Christian Approach to Animals," in Paul Waldau and Kimberly Patton, eds., *A Communion of Subjects* (New York: Columbia University Press, 2006), 143.

26. Winkler, *Magic of the Ordinary*, 72, 169.

27. Nee, *Spiritual Man*, 62. Nee states (page 45), "The knowledge of good and evil in this world is itself evil." This implies that duality is evil.

28. Levine, *Old Testament*, 8.

29. Quotes from now-defunct website www.askarabbi.com; Rabbi Dan posted on July 14, 2005; Rabbi Finman posted on October 20, 2000; and Rabbi Ben-Meir posted on December 6, 2000; accessed on September 23, 2005.

30. Winkler, *Magic of the Ordinary*, 169.

31. William R. Cook and Ronald B. Herzman, "Preaching and Ministries of Compassion," lecture 7, *Francis of Assisi* (Chantilly, VA: The Teaching Company, 2000), audiocassette.

32. Cook and Herzman, Lecture 8, "Knowing and Experiencing Christ," *Francis of Assisi*.

33. Andrew Linzey, *Animal Rites: Liturgies of Animal Care* (Cleveland: Pilgrim Press, 2001), 2.

34. Jimmy Akin, *National Catholic Register*, "Did Pope Francis Say Animals Go to Heaven?" December 13, 2014, http://www.ncregister.com/blog/jimmy-akin/did-pope-francis-say-animals-go-to-heaven.

35. Shanna Johnson, "A Heaven for All: Pope Francis Has Finally Laid to Rest the Debate Over Whether or Not Animals Can Go to Heaven," *US Catholic*, February 11, 2016, http://www.uscatholic.org/blog/201602/heaven-all-30553.

36. Linzey, *Animal Rites*, 5.

37. Randy Alcorn, *Heaven* (Wheaton, IL: Tyndale House Publishers, 2004), 381.

38. Bart D. Ehrman, "John — Jesus the Man from Heaven," lecture 8, *The New Testament*.

39. Rabbi Roller quoted on now-defunct website www.askarabbi.com; posted August 19, 1999, accessed September 23, 2005.

40. Winkler, *Magic of the Ordinary*, 47.

41. Cooper, "Prioritizing Your Life," session 1, *Holy Chariot*.

42. Bowker, *Oxford Dictionary*, 347.

43. Luke Timothy Johnson, "What's So Special about Matthew, Mark, Luke, and John?" *Jesus and the Gospels* (Chantilly, VA: The Teaching Company, February 2007), 9.

44. Alcorn, *Heaven*, 383.

Chapter 8. Science: Seeking Evidence

1. Robert M. Hazen, "The Mill-Urey Experiment," lecture 9, *Origins of Life* (Chantilly, VA: The Teaching Company, 2005), audiocassette.

2. Mamta Patel Nagaraja, "Dark Energy, Dark Matter," NASA, accessed May 28, 2017, https://science.nasa.gov/astrophysics/focus-areas/what-is-dark-energy.

3. Steven L. Goldman, "From Equilibrium to Dynamism," lecture 17, *Science in the 20th Century: A Social-Intellectual Survey* (Chantilly, VA: The Teaching Company, 2006), audiocassette.

4. Steven L. Goldman, "Isaac Newton's Theory of the Universe," lecture 4, *Science Wars: What Scientists Know and How They Know It* (Chantilly, VA: The Teaching Company, 2006), audiocassette.

5. Richard Wolfson, "A Problem of Gravity," lecture 13, *Einstein's Relativity and the Quantum Revolution: Modern Physics of Non-scientists* (Chantilly, VA: The Teaching Company, 2000), audiocassette.

6. Goldman, "Truth, History, and Citizenship," lecture 24, *Science Wars*.

7. Ibid.

8. Goldman, "Intelligent Design and the Scope of Science," lecture 23, *Science Wars*.

9. Lawrence M. Principe, "Science and Religion," lecture 1, *Science and Religion* (Chantilly, VA: The Teaching Company: 2006), audiocassette.

10. S. James Gates, Jr., "The Macro/Micro/Mathematical Connection," lecture 1, *Superstring Theory: The DNA of Reality* (Chantilly, VA: The Teaching Company, 2006), DVD.

11. Wolfson, "Uncommon Sense — Stretching Time," lecture 8; "Muons and Time-Traveling Twins," lecture 9; and "Escaping Contradiction — Simultaneity Is Relative," lesson 10, *Einstein's Relativity*.

12. Irene M. Pepperberg, *Alex and Me* (New York: Harper, 2008); Bernd Heinrich, *Mind of the Raven* (New York: Harper Perennial, 2006).

13. Edward A. Wasserman and Thomas R. Zentall, eds., *Comparative Cognition: Experiment Explorations of Animal Intelligence* (Oxford: Oxford University Press, 2006), Frans De Wal, *Are We Smart Enough to Know How Smart Animals Are?* (New York: W. W. Norton, 2016); Carl Safina, *Beyond Words: What Animals Think and Feel* (New York: Henry Holt, 2015).

14. Safina, *Beyond Words*, 27–29.

15. Frederick Gregory, "Consolidating Newton's Achievement," lecture 2, *The History of Science: 1700–1900* (Chantilly, VA: The Teaching Company, 2006), audiocassette.

16. Quote by John William Dawson from *The Story of the Earth and Man* (1887), vi, https://todayinsci.com/D/Dawson_John/DawsonJohn-Quotations.htm.

17. Gates, "Tying Up the Tachyon Monster with Spinning Strings," lecture 10, *Superstring Theory*.

18. Quote by Carl Sagan from http://www.quotationspage.com/quote/30374.html.

19. Goldman, "Locke, Hume, and the Path to Skepticism," lecture 6, *Science Wars*.

20. Gates, "The Macro/Micro/Mathematical," *Superstring Theory*.

21. For a list of renowned scientists who disagree with the big bang theory, see "Big Bang Theory: An Overview," All About Science, http://www.big-bang-theory.com.

22. Hazen, "The Grand Question of Life's Origin," lecture 1, *Origins of Life*.

23. Hazen, "Life on Clay, Clay as Life," lecture 17, *Origins of Life* .

24. Hazen, "Deep Space Dust, Molten Rock, and Zeolite," lecture 14, *Origins of Life*.

25. Aislinn Laing, "Four-Legged Animals Walked on Earth '18 Million Years Earlier Than Previously Thought,'" *Telegraph*, January 11, 2010, accessed May 28, 2017, http://www.telegraph.co.uk/news/science/science-news/6945194/Four-legged -animals-walked-on-earth-18-million-years-earlier-than-previously-thought.html. See also Kate Wong, "Ancient Stone Tools Force Rethinking of Human Origins," *Scientific American*, May 2017, https://www.scientificamerican.com/article/ancient -stone-tools-force-rethinking-of-human-origins. This article also announced the discovery of a 3.3-million-year-old stone tool that overturns longstanding views on human evolution.

26. Temple Grandin, *Animals in Translation* (New York: Scribner, 2005), 52–54.

27. Christof Koch, "Playing the Body Electric," *Scientific American Mind* (March/April 2010), 18–19.

28. Daniel N. Robinson, "Consciousness and the End of Mental Life," lecture 12, *Consciousness and Its Implications* (Chantilly, VA: The Teaching Company, 2007), audiocassette.

29. Jeffery Long, *Evidence of the Afterlife* (New York: Harper, 2010).

30. Eben Alexander, *Proof of Heaven: A Neurosurgeon's Journey into the Afterlife* (New York: Simon & Schuster, 2012).

31. Christian de Quincey, *Radical Nature: Rediscovering the Soul of Matter* (Montpelier, VT: Invisible Cities Press, 2002), 45–47.

32. Ibid., 49. De Quincey is a professor of philosophy and consciousness studies at John F. Kennedy University, is managing editor of *IONS Review* from the Institute of Noetic Sciences, and has published articles in numerous academic journals, including *Journal of Consciousness Studies*, *ReVision*, *Network*, *Cerebrum*, and *Journal of Transpersonal Psychology*.

33. Bernardo Kastrup, *Why Materialism Is Baloney* (Washington, DC: Iff Books, 2014), 66–67.

34. Marc D. Hauser, *Wild Minds: What Animals Really Think* (New York: Henry Holt and Company, 2001), xviii, 102.

35. Joshua M. Plotnik, Frans de Waal, and Diana Reiss, "Self-Recognition in an Asian Elephant," *Proceedings of the National Academy of Sciences of the United States of America* 103, no. 45 (September 13, 2006), doi: 10.1073/pnas.0608062103.

36. Diana Reiss and Lori Marino, "Mirror Self-Recognition in the Bottlenose Dolphin: A

Case of Cognitive Convergence," *Proceedings of the National Academy of Sciences of the United States of America* 98, no. 10 (October 3, 2000), doi: 10.1073 /pnas.101086398.

37. Bernardo Kastrup, *More Than Allegory: On Religious Myth, Truth and Belief* (Washington, DC: Iff Books, 2012–15).

38. Robinson, *Consciousness and Its Implications*.

39. Rhawn Joseph, "Limbic System: Sex, Hallucinations, Emotions, Memory, PTSD, Amygdala...Brain Mind, Lecture 6," YouTube.com, posted December 9, 2006, https://www.youtube.com/watch?v=T7nXiXQb2iM.

40. Neel Burton, "What Are Basic Emotions?" *Psychology Today*, January 7, 2016, https://www.psychologytoday.com/blog/hide-and-seek/201601/what-are-basic -emotions.

41. Heinrich, *Mind of the Raven*, 336.

42. Ibid., 313, 323.

43. Ibid., 341.

44. Wasserman and Zentall, *Comparative Cognition*, 9.

45. Irene M. Pepperberg, *Alex and Me* (New York: Harper, 2009), frequently asked questions, 107–11.

46. See *Current Biology* 19, no. 21 (November 17, 2009).

47. Pepperberg, *Alex and Me*, 94–95.

48. Ibid., 150.

49. See *The Noetic Post: A Bulletin from the Institute of Noetic Sciences* 3, no. 2 (spring/ summer 2012).

50. "The Cambridge Declaration on Consciousness," Francis Crick Memorial Conference, July 7, 2012, http://fcmconference.org/img/CambridgeDeclarationOn Consciousness.pdf.

51. Barbara J. King, "Tool Making — Of Hammers and Anvils," lecture 6, *Roots of Human Behavior* (Chantilly, VA: The Teaching Company, 2001), audiocassette.

52. Grandin, *Animals*, 255.

53. Ibid., 254–60.

54. Matt Walker, "Burrowing US Prairie Dogs Use Complex Language," *BBC Earth News*, February 2, 2010. http://news.bbc.co.uk/earth/hi/earth_news/newsid_8493000 /8493089.stm.

55. Grandin, *Animals*, 273–75.

56. Edward J. Larson, *The Theory of Evolution: A History of Controversy* (Chantilly, VA: The Teaching Company, 2002), lecture 12, p. 51.

57. Hauser, *Wild Minds*, 239.

58. Marc Bekoff, "Wild Justice, Social Cognition, Fairness, and Morality," in Paul Waldau and Kimberley Patton, eds., *A Communion of Subjects: Animals in Religion, Science, and Ethics* (New York: Columbia University Press, 2006), 463–64.

59. Sathya Sai Baba quoted in interview with R. K. Karanjia, "Swama Explains — The Blitz Interview," *Blitz News Magazine* (September 1976), http://www.saibaba.ws /articles2/blitz.htm.

Chapter 9. Spirituality: Mystics, Clairvoyants, Channels, and Animal Communicators

1. Eckhart Tolle and Patrick McDonnell, *Guardians of Being* (Novato, CA: New World Library, 2009), 106.

2. Larry Dossey, *Healing Words: The Power of Prayer and the Practice of Medicine* (San Francisco: HarperSanFrancisco, 1993). Also quoted in Renée A. Scheltema, *Something Unknown Is Doing We Don't Know What* (Telekan, 2009), DVD.

3. Eckhart Tolle, *The Power of Now* (Novato, CA: New World Library, 1999), 164.

4. Abraham-Hicks, "5/19/04 Buffalo, New York, Complete Workshop Recording" (San Antonio, TX: Abraham-Hicks Publications, 2004), audiocassette.

5. Tolle, *Power of Now*, 165–66.

6. Ibid., 157.

7. *Species Link: The Journal of Interspecies Telepathic Communication*, no. 83 (summer 2011), 18–22.

8. Tolle and McDonnell, *Guardians*, 52.

9. Melissa Healy, "Here's Scientific Proof That Your Dog Feels You," *Los Angeles Times*, January 12, 2016, http://www.latimes.com/science/la-sci-sn-your-dog-feels-you -20160112-story.html.

10. Mark Denicola, "Study Reveals Heartbeats of Owners and Dogs Synch Up When United," Collective Evolution, May 10, 2016, http://www.collective-evolution.com /2016/05/10/study-reveals-heartbeats-of-owners-dogs-synch-up-when-united -video.

11. For more information on how meditation affects the brain, see Andrew Newberg and Mark Robert Waldman, *How God Changes Your Brain* (New York: Random House, 2009).

12. David R. Hawkins, *Discovery of the Presence of God* (Sedona, AZ: Veritus Publishing, 2007), 115.

13. Jill Bolte Taylor, *My Stroke of Insight: A Brain Scientist's Personal Journey* (New York: Penguin Group, 2006), 3, 20, 79, 140.

14. Ibid., 71.

15. David R. Hawkins, *Transcending the Levels of Consciousness: The Stairway to Enlightenment* (Sedona, AZ: Veritus Publishing: 2006), xvii.

16. Ibid., xvii–xix.

17. Ibid., 40.

18. Ibid., 28–29, 31, 167, 250.

19. Ibid., 323.

20. David R. Hawkins, *The Eye of the I from Which Nothing Is Hidden* (Sedona, AZ: Veritus Publishing, 2002), 228.

21. John Bowker, ed., *The Oxford Dictionary of World Religions* (Oxford: Oxford University Press, 1997), 519.

22. Gershon Winkler, *Kabbalah 365: Daily Fruit from the Tree of Life* (Kansas City: Andrews McMeel Publishing, 2004), 57, 236. In the Bible, this idea appears in Psalms 8:7–8; 145:10; 148:3–4 and 7–11; Isaiah 55:12; Job 12:7–8; and also in *Midrash Heichalot Rabati* 24:3.

23. Bowker, *Oxford Dictionary*, 106, 107.

24. Kabir Helminski, ed., *The Pocket Rumi* (Boston: Shambhala, 2008), 150.

25. Bowker, *Oxford Dictionary*, 925–26.

26. Michael Bernard Beckwith, *40 Day Mind Fast Soul Feast* (Los Angeles: Agape Media International, 2007), Day 12.

27. Thomas Merton, *Choosing to Love the World* (Boulder, CO: Sounds True, 2008), 102.

A Happy Ending

1. Sogyal Rinpoche, *Tibetan Book*, 11.

2. Ibid., 256, 303.

3. Ibid., 301–2.

4. Bancroft, *Buddha Speaks*, 8.

5. Sogyal Rinpoche, *Tibetan Book*, 134.

6. Lorie Eve Dechar, *The Five Spirits: Alchemical Acupuncture for Psychological and Spiritual Healing* (New York: Chiron Publications, 2006), 59.

7. Sogyal Rinpoche, *Tibetan Book*, 292.

8. Ibid., 224.

Index

Cordyceps mushroom, 99, 202

Cosmic Serpent, The, 193

Couch, Stacy, 54

Course in Miracles, A, 215

cowbirds, 221–22

cowboy poetry, 10–11

cows, 126

Crazy Horse (Sioux chief), 55

creation myths, 30–31; African tribal, 39–41; animal symbols in, 41–42; Australian Aborigine, 35–37, 104; biblical, 37–39, 176–78; common themes in, 31–32; controversies over, 31; Hopi, 32–34; science and, 200–202

creativity, 41, 42

Crete, 94

crows, 126

crystals, 60

Cushing's disease, 186

daemon, 55

dakinis, 158–59

Dalai Lama, 136, 137, 138, 142–43, 153, 154

Dan, Rabbi, 184

Dana (author's clairvoyant friend). *See* Xavier, Dana

Darwin, Charles, 201, 210

Dawn (author's client), 18–21, 74

Dawson, John William, 200

death: of animals, La Plata County (CO) examples, 243–52; animal symbols for, 42; Buddhist view of, 151, 165–66, 242–43, 247–48; energy at time of, 200; euthanasia, 42–47; good transitions to, 242–43; pagan

views of, 92–93, 102; of pets, 124–25; prayers after, 247; shamanism and, 49–50; spiritual concerns about, 241–42; veterinarians and, 4

Dee (author's friend), 222–23, 224

deer, 94

demonic art, 154

demons, 153, 158–59

Denise (author's friend), 212

Denver (ER doctor), 120

Derinkuyu (Turkey), 33

Deuteronomy, Book of, 102

devil, 176

dharma, 148

disease, 75–76

DNA, 36–37, 42, 98, 101–2, 193, 201, 202

dogs: as conscious, 238; deaths of, 8–10, 241; domestication of, 70, 94; euthanizing, 90; human communication with, 227–31, 232, 234, 235; karma and, 114; love demonstrated by, 238; moral judgment capacity of, 16, 214; in pagan rituals, 107–8; spirit realm awareness of, 159; veterinary care of, 22, 26, 104

dolphins, 206

domestication, 70, 94, 103

donkeys, 180–82

Dossey, Larry, 216

doves, 191, 192

Dowden, Ken, 78, 108

dragons, 100, 101–2, 154, 158, 196

Dravidian civilization, 111

dreams, 216

Dreamtime, 35, 79, 104

Druids, 78, 105

About the Author

KARLENE STANGE, DVM, as a child called herself an "animal doctor" before she knew the word "veterinarian." Today she incorporates acupuncture, traditional Chinese medicine, and herbal and nutritional therapy into her rural Rocky Mountain practice. She often speaks at conferences and lives in Durango, Colorado. You can learn more at www.AnimasAnimals.com.

NEW WORLD LIBRARY is dedicated to publishing books and other media that inspire and challenge us to improve the quality of our lives and the world.

We are a socially and environmentally aware company. We recognize that we have an ethical responsibility to our customers, our staff members, and our planet.

We serve our customers by creating the finest publications possible on personal growth, creativity, spirituality, wellness, and other areas of emerging importance. We serve New World Library employees with generous benefits, significant profit sharing, and constant encouragement to pursue their most expansive dreams.

As a member of the Green Press Initiative, we print an increasing number of books with soy-based ink on 100 percent postconsumer-waste recycled paper. Also, we power our offices with solar energy and contribute to nonprofit organizations working to make the world a better place for us all.

Our products are available in bookstores everywhere.

www.newworldlibrary.com

At NewWorldLibrary.com you can download our catalog,
subscribe to our e-newsletter, read our blog,
and link to authors' websites, videos, and podcasts.

Find us on Facebook, follow us on Twitter, and watch us on YouTube.

Send your questions and comments our way!
You make it possible for us to do what we love to do.

Phone: 415-884-2100 or 800-972-6657
Catalog requests: Ext. 10 | Orders: Ext. 10 | Fax: 415-884-2199
escort@newworldlibrary.com

NEW WORLD LIBRARY
publishing books that change lives   14 Pamaron Way, Novato, CA 94949